Chaos Theory
Finding Order in the Universe's Chaos

Copyright © 2024 by Nobtrex LLC

All rights reserved. No part of this publication may be reproduced, distributed, or transmitted in any form or by any means, including photocopying, recording, or other electronic or mechanical methods, without the prior written permission of the publisher, except in the case of brief quotations embodied in critical reviews and certain other noncommercial uses permitted by copyright law.

Contents

1 **Introduction to Chaos Theory** — 11
   1.1 What is Chaos Theory? 11
   1.2 Historical Overview of Chaos Theory . . . 14
   1.3 Fundamental Concepts and Terms 16
   1.4 The Scope of Chaos Theory 19
   1.5 Misconceptions and Clarifications 21
   1.6 Real-World Examples of Chaos 23
   1.7 Importance of Studying Chaos Theory . . . 25

2 **The Butterfly Effect: Sensitive Dependence on Initial Conditions** — 29
   2.1 Defining the Butterfly Effect 29
   2.2 Origins and Development of the Concept . 32
   2.3 Mathematical Foundations of the Butterfly Effect 34
   2.4 Examples in Meteorology 36
   2.5 Experimental Observations and Evidence . 39
   2.6 Butterfly Effect in Other Systems 41

   2.7   Implications for Predictive Models 44
    2.8   The Butterfly Effect in Popular Culture . . . 46

3 Fractals: The Patterns of Chaos   49
    3.1   Introduction to Fractals 49
    3.2   Properties of Fractals 51
    3.3   Generating Fractals: Methods and Techniques. 53
    3.4   Famous Fractals and Their Characteristics . 56
    3.5   Fractals in Nature 58
    3.6   Applications of Fractals in Science and Technology 60
    3.7   Fractals in Art and Design 63
    3.8   Mathematical Challenges and Advances in Fractal Theory 65

4 Nonlinear Dynamics: Understanding the Rules 69
    4.1   Introduction to Nonlinear Dynamics 69
    4.2   Contrasting Linear and Nonlinear Systems 71
    4.3   Key Principles of Nonlinearity 74
    4.4   Tools for Analyzing Nonlinear Dynamics . 76
    4.5   Stability and Instability in Systems 79
    4.6   Bifurcations: Changes in System Behavior . 81
    4.7   Chaos in Nonlinear Systems 84
    4.8   Case Studies: Nonlinear Dynamics in Action 86

5 Chaos in Nature: Weather, Ecology, and Beyond 91

CONTENTS

    5.1   Chaos in Weather Patterns 91

    5.2   Ecological Systems and Chaos 93

    5.3   Geophysical Phenomena and Nonlinear Dynamics 96

    5.4   Biological Chaos: From Heartbeats to Neural Networks 98

    5.5   Impact of Chaos Theory on Environmental Science 101

    5.6   Predicting and Managing Natural Disasters 103

    5.7   Long-term Implications for Climate Change 105

6 Chaos in Technology: From Internet Traffic to Stock Markets   109

    6.1   Overview of Chaos in Technological Systems 110

    6.2   Chaos in Communication Networks 112

    6.3   Internet Traffic and Nonlinear Dynamics . . 114

    6.4   Stock Market Fluctuations and Predictive Models 117

    6.5   Chaos in Electronic Circuits and Devices . . 119

    6.6   Applications in Cryptography and Security 122

    6.7   Adaptive Technologies Harnessing Chaos . 124

    6.8   Challenges and Future Directions in Technological Chaos 127

7 Predicting the Unpredictable: Limitations and Possibilities   131

    7.1   The Paradox of Predicting Chaos 132

- 7.2 Methods of Forecasting Chaotic Systems . . 134
- 7.3 Uncertainty and Errors in Predictions . . . 136
- 7.4 Limitations of Current Models 138
- 7.5 Advances in Computational Techniques . . 140
- 7.6 Machine Learning and AI in Prediction . . 142
- 7.7 Real-world Case Studies 145
- 7.8 Ethical Considerations and Responsible Forecasting 148

8 Quantum Chaos: Bridging Micro and Macro Scales   151

- 8.1 Introduction to Quantum Chaos 151
- 8.2 Foundational Concepts in Quantum Mechanics and Chaos Theory 154
- 8.3 Differences and Similarities with Classical Chaos 157
- 8.4 Tools and Techniques for Analyzing Quantum Chaos 159
- 8.5 Quantum Chaos in Atomic and Molecular Physics 161
- 8.6 Applications in Quantum Computing . . . 163
- 8.7 Impact on the Understanding of Quantum Decoherence 165
- 8.8 Future Research Directions in Quantum Chaos 168

9 Chaos in Art and Culture: Influence and Inspiration   171

- 9.1 Exploring the Connection Between Chaos Theory and Art 171
- 9.2 Visual Arts: Fractals and Patterns 174
- 9.3 Music and Chaos: Compositional Techniques 176
- 9.4 Literature: Narratives and Themes of Chaos 179
- 9.5 Architecture: Incorporating Nonlinear Dynamics 181
- 9.6 Cinema: Depicting Chaos and Complexity 183
- 9.7 Cultural Impacts: Perceptions of Chaos in Society 185
- 9.8 Artistic Expression as a Reflection of Chaotic Concepts 188

10 The Future of Chaos: Emerging Trends and Theories — 193

- 10.1 Current State of Chaos Theory Research . . 194
- 10.2 Emerging Technologies and Their Impact on Chaos Theory 196
- 10.3 Integration of Chaos Theory Across Disciplines 198
- 10.4 New Mathematical Tools and Approaches . 201
- 10.5 Future Applications in Science and Engineering 203
- 10.6 Evolving Theories in Quantum Chaos . . . 206
- 10.7 Challenges and Opportunities for Innovation 208
- 10.8 Ethics and Implications of Advanced Chaotic Systems 211

CONTENTS

Preface

Chaos Theory is a fascinating and intricate field that lurks in the realms of mathematics and science, revealing patterns, structures, and phenomena that initially appear wholly random and undirected. This book, "Chaos Theory: Finding Order in the Universe's Chaos," is aimed at demystifying the complex ideas and principles of chaos theory for an educated but non-specialist audience. With its profound implications across various disciplines such as physics, biology, economics, and even philosophy, an understanding of chaos theory can enrich one's perspective on the world.

The objectives of this book are multifold. Firstly, it seeks to provide a clear and concise introduction to chaos theory, explaining its key concepts and the mathematical underpinnings without overwhelming the reader with technical details. Additionally, the book aims to illustrate the broad applicability of chaos theory through examples and case studies that highlight its relevance in real-world scenarios. Through this approach, the text endeavors to not only educate but also inspire readers by showcasing the versatile utility and often unexpected beauty of chaotic systems.

The substance of the book is divided into chapters that each tackle distinct aspects of chaos theory, starting with foundational concepts and gradually moving towards more advanced topics and applications. From the iconic 'butterfly effect' to the intricate patterns of fractals, each chapter is structured to build upon previous knowledge while introducing new information in a digestible format.

This book is designed for readers who possess a college-level education. It is ideal for individuals with a background or keen interest in science, technology, engineering, and mathematics (STEM), as well as educators and professionals looking to broaden their understanding of this pivotal theory. Additionally, the book is an excellent resource for anyone involved in fields where chaos theory's ideas have made a significant impact, such as economics, meteorology, and environmental science.

In summary, "Chaos Theory: Finding Order in the Universe's Chaos" aspires to be a comprehensive yet accessible text that will equip readers with a robust understanding of chaos theory, enabling them to appreciate the intricacies and applications of this dynamic field. Through this book, readers are invited to explore the orderly patterns hidden within the chaos that surrounds us.

Chapter 1

Introduction to Chaos Theory

Chaos theory challenges traditional notions of predictability and order in systems, ranging from weather patterns to economic markets. This field of study focuses on understanding how and why seemingly random states in complex systems develop and evolve. The introduction provides a foundational overview, discussing the historical background, key concepts, and terminology essential for a grasp of chaos theory, making it approachable for readers with a general interest in science and mathematics.

1.1 What is Chaos Theory?

Chaos Theory is a mathematical framework that investigates the behavior of dynamical systems that are highly sensitive to initial conditions, a phenomenon popularly referred to as the butterfly effect. This branch of mathematics describes systems so sensitive that small differences in the initial conditions can lead to vastly different outcomes, making long-term predictions almost impossible with high accuracy.

Contrary to its name, Chaos Theory does not imply a lack of order. Rather, it explores how underlying patterns, nuanced structures, and deterministic laws can lead to apparently unpredictable and erratic behaviors in complex systems. It occupies the intersection of mathematics, physics, engineering, economics, biology, and several other fields, illustrating its broad applicability and profound influence.

The foundational element in chaos theory is the dynamical system. These systems are defined by a set of rules or equations that describe how points in a given space evolve over time. These could be the positions of planets, the price of a stock, the weather, and so forth. Often represented mathematically by differential equations or iterative maps, these models establish the environment within which chaos can manifest.

Chaos is typically recognized by the following characteristics:

1. **Sensitivity to initial conditions:** As noted earlier, this is often characterized by the popular metaphor of the butterfly effect. A butterfly flapping its wings in Brazil could potentially cause a tornado in Texas. Mathematically, this means that even minuscule changes in initial conditions within a chaotic system lead to diverging outcomes, making precise predictions virtually impossible over significant time periods.

2. **Topological mixing:** This property of chaotic systems means that the system will evolve over time such that

1.1. WHAT IS CHAOS THEORY?

any given region or open set within its phase space eventually overlaps with any other given region. Essentially, variables within the system get so thoroughly mixed over time that their individual origins become indistinguishable.

3. **Dense periodic orbits:** In a chaotic system, there are countless periodic orbits, and these are densely woven such that they come arbitrarily close to any point in the system's phase space.

To demonstrate the concept of chaos, consider a simple mathematical example such as the logistic map, defined by the recurrence relation:

$$x_{n+1} = rx_n(1 - x_n).$$

Here, x represents a population at any stepped time n, and r is a positive constant that represents the rate of growth. As you adjust r, the behavior of the map dramatically changes from stable to chaotic, showcasing sensitive dependence on parameters — another characteristic aspect of chaotic systems.

Although chaos theory originated from observing weather systems and trying to predict weather patterns, Edward Lorenz's seminal work in the 1960s highlighted its broad implications. The profound realization was that even a system with deterministic rules could behave unpredictably in practical terms due to our inability to measure initial conditions with infinite precision.

In visual terms, chaos can be illustrated using phase diagrams and bifurcation diagrams that visually represent how dynamical systems change and evolve under varying initial conditions or parameters. These diagrams are incredibly rich with intricate structures that are characteristic of chaotic systems — fractual boundaries between basins of attraction of different attractors, spiraling patterns indicating sensitivity to initial conditions, and branches demonstrating how slight changes in parameters can lead to dramatic shifts in system behavior.

As intriguing as it might be conceptually, chaos presents challenges and opportunities. It pushes the boundaries of how we understand predictability and control in systems as varied as cardiac rhythms in medicine, stock market behaviors in economics, and climate change in environmental science. Through this exploration of chaos, we develop new mathematical tools and insights that translate directly into better-informed decision-making across numerous fields.

By venturing deeper into chaos theory, we uncover more about the interconnectedness and intricate dependencies that shape complex behaviors in both nature and human-made systems. It invites us to rethink our approaches to understanding and navigating complexity and unpredictability in our world.

1.2 Historical Overview of Chaos Theory

Chaos theory, although it seems a product of mid-20th century scientific advancements, has roots extending much further back. The journey to formalize chaos theory began subtly through observations and calculations in mathematical astronomy and meteorology, where the prediction limitations due to complex dynamic behavior first became apparent.

The conceptual seed of chaos theory can arguably be

1.2. HISTORICAL OVERVIEW OF CHAOS THEORY

traced back to Henri Poincaré in the late 19th and early 20th centuries. Poincaré, a French mathematician often lauded for his work on the behavior of the three-body problem in celestial mechanics, provided the earliest thoughts resembling chaos theory. His work demonstrated that even simple systems could exhibit unpredictable and erratic behavior due to sensitivity to initial conditions, later recognized as a hallmark of chaotic systems.

Fast forward to the mid-20th century, a significant contributor, Edward Lorenz, an American mathematician and meteorologist, made profound advancements that propelled chaos theory into a formal field of study. In 1963, Lorenz developed a simplified mathematical model for atmospheric convection, consisting of a set of differential equations. These modeled equations unexpectedly displayed non-repeating, non-linear dynamics when small changes in initial conditions were applied, leading to what is now famously known as the "butterfly effect."

Following Lorenz's discoveries, there was a wave of interest in irregular, non-linear dynamics across various scientific disciplines. During the 1970s and 1980s, scientists such as Mitchell Feigenbaum contributed extensively by discovering universal properties in chaotic systems related to scaling and iteration through his work with the logistic map. This era also saw the development of the Mandelbrot set by Benoît Mandelbrot, which provided a geometric visualization of iterative complex functions, where the boundary's intricate structure embodies chaotic dynamics.

The field's expansion continued with the development of tools and methods to analyze chaotic systems, notably by James Yorke and T. Y. Li who coined the term 'chaos' in a

1975 paper discussing the periodicity and unpredictability in certain dynamical systems. These methods included Lyapunov exponents and fractal dimensions that measure the rate of separation of infinitesimally close trajectories and the complex geometric scales of chaotic attractors, respectively.

The interdisciplinary nature of chaos theory became evident as its principles were recognized and utilized across physics, chemistry, biology, economics, and even philosophy. Each discipline adapted the core concepts of chaos theory to explore complex systems within their specific domains, establishing chaos theory not just as a mathematical curiosity but as a fundamental lens through which the unpredictable becomes somewhat predictable.

Through revisiting these historical milestones, we appreciate how chaos theory has evolved from speculative observation to foundational science. Its influence spans across predicting weather systems to understanding population growth dynamics, ultimately illustrating an ongoing narrative of intricate connections that dictate the fabric of numerous natural and human-made systems. Let us continue exploring how these principles apply in diverse scenarios and discover what further insights await in the unpredictable yet fascinating realm of chaos.

1.3 Fundamental Concepts and Terms

As we dive deeper into the intricacies of chaos theory, it becomes imperative to familiarize ourselves with the fundamental concepts and terms that form the backbone of this intriguing field. This section aims to meticulously unpack these terms, providing a thorough understanding that will aid in decoding the complexities of chaotic systems.

Dynamical Systems: At the core of chaos theory lies the concept of dynamical systems. These systems can be described as mathematical constructs used to model processes that evolve over time according to a specific set of rules. These can be as straightforward as the equations describing the swinging of a pendulum or as intricate as those predicting the weather.

Attractors: An attractor is a set towards which a dynamical system evolves after a long enough period. For instance, when observing the weather as a dynamical system, an attractor could represent a state wherein the system tends to show repetitive behavior like seasonal cycles. However, in chaotic systems, these attractors can have complex structures known as strange attractors.

Strange Attractors: These are attractors that exhibit a fractal structure, meaning they are self-similar across different scales. Strange attractors are a hallmark of chaotic systems. The Lorentz attractor, discovered by Edward Lorenz, is a famous example, often visualized as a butterfly-like structure demonstrating sensitive dependence on initial conditions.

Initial Conditions: In chaos theory, initial conditions refer to the state of the system at the start of observation. Small differences in initial conditions can lead to vastly different outcomes, which is popularly known as the butterfly effect.

Term	Definition
Dynamical Systems	Mathematical models that describe processes evolving over time
Attractors	States towards which systems evolve and stabilize over time
Strange Attractors	Fractal structures toward which chaotic systems evolve
Initial Conditions	The starting state of a system, crucial in chaos theory for predicting outcomes

Sensitive Dependence on Initial Conditions: Often referred to as the butterfly effect, this concept is central to understanding chaos. It describes how tiny changes in the initial state of a system can lead to significant and unpredictable changes in its future state. The name "butterfly effect" comes from the metaphorical example that the flap of a butterfly's wings in Brazil could set off a tornado in Texas.

Utilizing these foundational concepts, chaos theory helps us understand the unpredictability inherent in complex systems, despite deterministic laws. The field encourages looking beyond initial appearances to uncover underlying patterns and behaviors, a reminder that what appears chaotic may not be entirely bereft of order.

1.4 The Scope of Chaos Theory

Chaos theory, often perceived as the study of disorder, finds its true essence in uncovering the underlying order within complex and seemingly unpredictable systems. This scope is not limited to any one discipline but spans across various fields, each offering critical insights into the intricate dance between determinism and randomness.

To begin, let's delve into the realm of meteorology, where chaos theory first made its mark with Edward Lorenz's discovery of the butterfly effect. In this context, tiny variations in the initial state of the weather system can lead to vastly different outcomes, which makes long-term weather forecasting an immense challenge. This sensitivity to initial conditions is a quintessential example of chaotic behavior in a natural system and underscores the necessity for highly accurate data in weather prediction systems.

Further extending its reach, chaos theory is pivotal in understanding the dynamics within ecological systems. Populations of species, for instance, often exhibit chaotic behavior due to intricate dependencies and interactions with their environment. Models considering these factors can help in predicting sudden population explosions or crashes, assisting in conservation efforts and understanding natural cycles more thoroughly.

In the field of engineering, chaos plays a significant role in the study of nonlinear dynamical systems, such as bridges, skyscrapers, and even spacecraft. Engineers model these structures to predict how they will respond to various stresses, including those from environmental factors. A minor oversight in predicting chaotic responses can lead to catastrophic failures, demonstrating the critical application of chaos theory in safety analyses and design robustness.

Moving to the realm of medicine, researchers apply chaos theory to understand complex systems within the human body. For example, the rhythm of the human heart and the brain's neural network display chaotic characteristics that can sometimes lead to irregularities or disorders. By applying principles of chaos theory, medical professionals can better predict and manage conditions such as cardiac arrhythmias or epilepsy.

In economics, chaos theory helps in scrutinizing market dynamics which are influenced by numerous intertwined factors leading to apparently stochastic fluctuations in markets like stock prices and commodities. Economists leveraging chaos theory attempt to unearth patterns that traditional models might overlook, aiming for more accurate predictions and risk assessments.

The scope of chaos theory also stretches into technology, notably in the realm of algorithm development and cryptography. The unpredictable nature of chaotic systems makes them ideal candidates for developing secure encryption techniques for data transmission. On a similar note, algorithms that emulate chaotic dynamics can assist in optimizing solutions to complex logistical or operational problems in business and industry.

Lastly, in the arts and humanities, chaos theory influences new perspectives in understanding human behavior, linguistic evolution, and even philosophical constructs regarding free will and determinism. Artists and writers draw upon chaos theory to explore themes of complexity and unpredictability inherent in human existence.

While exploring such a vast array of applications, it becomes increasingly evident that chaos theory is not merely a scientific curiosity but a fundamental lens through which to view the complexities of systems across physical, biological, economic, and socio-cultural domains. Its capacity to reveal hidden structures within chaotic environments continues to inspire innovation across disciplines, prompting ongoing dialogue between theorists and practitioners about the constraints and potentials of understanding chaos.

1.5 Misconceptions and Clarifications

One of the most prevalent misconceptions about chaos theory is the belief that it implies complete randomness and lack of order in systems it describes. However, this could not be farther from the truth. Chaos theory, rather than describing systems as purely random, investigates how order and pattern can emerge from seemingly random behaviors in deterministic systems—those governed by fixed rules. This nuanced perspective is crucial for distinguishing chaos from randomness.

Another common misunderstanding is equating the chaotic nature of a system with it being unanalyzable or beyond the scope of scientific prediction. While it's true that the sensitivity to initial conditions, often referred to as the butterfly effect, poses significant challenges in predicting specific long-term outcomes in chaotic systems, this does not imply that chaotic systems are unmanageable. In fact, chaos theory provides tools to understand the broad behaviors and limits of these systems, despite their unpredictable specifics.

It's also important to dispel the notion that the study of chaotic systems is only applicable to grand-scale or abstract phenomena like weather systems or galaxy formation. On the contrary, chaos theory has practical applications in everyday technology and scientific fields. From the algorithms that stabilize electric power grids to the models optimizing inventory management in logistics, chaos theory finds relevance in various scales and scenarios.

Many also mistakenly think that chaos theory lacks mathematical rigor due to its association with unpredictable systems. Nevertheless, the analysis within chaos theory is deeply rooted in sophisticated mathematics, particularly in the areas of differential equations and dynamic systems. This profound mathematical foundation ensures that while outcomes cannot be precisely predicted, the characterizations of system behaviors can be deeply understood.

Thus, in understanding chaos, one must appreciate the complexity and subtleties of chaotic systems, noting both their unpredictability in specific outcomes and their overarching predictabilities in general behaviors. This duality is not only fascinating but also immensely valuable across various scientific disciplines.

To further explore this, consider the Lorenz attractor, a three-dimensional structure in phase space that helps visualize how chaotic dynamics evolve over time. The attractor, named after Edward Lorenz, one of the pioneers in chaos theory, illustrates how even in a deterministic system where you know all the rules, exact forecasting is still challenging due to the system's sensitive dependence on initial conditions. This visualization not only reinforces the deterministic nature of chaotic systems but also underscores the complexity within simplicity—an essential theme in chaos theory.

As we delve deeper into other sections, keep these clarifications in mind. They serve as a solid foundation for not only understanding the broader implications of chaos theory but also appreciating the intricate order that it brings to light from the apparent disorder. By recognizing these misconceptions and clarifications, we embrace a more accurate and enriched perspective on chaos theory, paving the way for more profound insights and innovative applications in our understanding of complex systems.

1.6 Real-World Examples of Chaos

Weather Systems: One of the most commonly cited examples of chaotic systems is the weather. Meteorologists describe weather using nonlinear equations to model how various atmospheric factors like temperature, pressure, and humidity interact. Edward Lorenz, a meteorologist and a key figure in chaos theory, found that tiny variations in initial conditions could greatly affect the long-term behavior of weather systems, a phenomenon widely known as the "butterfly effect." For instance, in simulations where initial temperature values differed by fractions of a degree, the resulting weather forecasts diverged dramatically over time. This sensitivity to initial conditions highlights why precise long-term weather forecasting becomes immensely challenging.

Population Dynamics: The study of population dynamics in ecology provides another showcase of chaos theory in a real-world setting. The populations of certain species can exhibit chaotic behavior based on intrinsic factors like reproduction rates and resource availability, as well as extrinsic factors such as climate variability and human intervention. A model that captures such chaotic dynamics is the logistic map, an equation that considers the reproduction rate and carrying capacity of an environment. As the reproduction rate increases, the population may enter a chaotic regime, where small changes in the reproduction rate or initial population can lead to significantly different outcomes, making long-term predictions unreliable.

Economic and Financial Systems: The markets are inherently unpredictable and often regarded as chaotic systems due to their complex interactions between myriad factors such as investor behavior, government policies, and external shocks. Stock market fluctuations can exhibit chaotic behavior where small events can disproportionately influence market movements. This unpredictability is why models that attempt to forecast economic outcomes can have limitations and require careful consideration of initial conditions and parameter values.

Heart Rhythms and Health Conditions: In medicine, the concept of chaos can be applied to understand certain health conditions, such as heart arrhythmias. The heartbeat may appear regular, but subtle nonlinear interactions among cells in the heart can lead to chaotic behavior, manifesting as irregular heart rhythms or medical conditions like ventricular fibrillation. Understanding this chaos helps medical professionals develop better diagnosis and treatment strategies that accommodate the sensitive nature of heart dynamics.

Traffic Flow: Traffic flow is another system where chaos theory applies. On highways, under certain conditions, slight changes in a driver's behavior (like sudden braking) can propagate through the traffic stream, leading to unpredictable stop-and-go patterns. Mathematical models considering the nonlinear interactions among vehicles provide insights that help in the design of traffic control systems that aim to minimize congestion and improve safety.

Throughout these examples, we observe how chaos theory provides pivotal insights that challenge traditional deterministic approaches, demonstrating its utility across diverse fields. The applicability of chaos theory to these dynamic systems is not merely academic but has practical implications for developing strategies to manage, predict, and optimize in complex environments. Understanding chaos invites us to respect the inherent unpredictability of systems and to rethink our approaches toward modeling and simulation, ensuring that we account for the profound impacts of minute differences in initial conditions. This rethinking allows scientists and researchers in various fields to devise more resilient systems and predictive models that better accommodate the natural variability and inherent unpredictabilities that characterize many aspects of our world.

1.7 Importance of Studying Chaos Theory

Chaos theory, as a field, serves as a monumental shift in how we understand natural and manmade systems. Its importance is underscored not solely by its ability to describe how systems evolve in unpredictable ways, but also in its profound implications across various scientific domains as well as everyday life applications.

One of the primary reasons for studying chaos theory lies in its potency to enhance prediction accuracy and decision-making in systems traditionally considered too complex or unpredictable. Consider meteorology, where chaos theory has revolutionized weather prediction models. By understanding that weather is a chaotic system with sensitive dependence on initial conditions, meteorologists can better comprehend and predict severe weather events by incorporating algorithms that consider chaotic factors into their models.

Another vital area where chaos theory finds its application is in the realm of medicine, particularly in cardiology and neuroscience. Irregular heart rhythms, like fibrillation, are chaotic in nature. Application of chaos theoretical approaches allows for better prediction and management of such conditions, potentially saving lives by foreseeing critical cardiac events before they occur. Similarly, in neuroscience, understanding the chaotic patterns in brain activity can lead to more effective strategies for treating conditions like epilepsy and other neurological disorders that exhibit chaotic dynamics.

The field of economics also benefits significantly from chaos theory. The unpredictable behavior evident in stock markets and economic systems can be better understood through the lens of chaos, providing insights that help predict market trends and prevent financial crises. Economists utilize chaos theory to model complex economic phenomena, helping to illuminate the underlying mechanics that drive market volatility and enabling better regulatory and policy decisions.

In engineering, chaos theory helps in designing more efficient and robust systems and technology. For instance, understanding chaotic vibrations in structures can lead

1.7. IMPORTANCE OF STUDYING CHAOS THEORY

to safer bridge designs and improved construction materials that are resilient to unpredictable stress loads. Similarly, in the fields of robotics and artificial intelligence, chaos theory is employed to develop systems that adapt freely to changing environments, enhancing automation technology across various industries including manufacturing, healthcare, and transportation.

Moreover, the study of chaos theory prompts a deeper understanding of the natural universe. It provides a framework for exploring the intricate patterns found in various natural processes, from the formation of galaxies in astrophysics to pattern formation in biological systems such as seashell growth and leaf arrangement.

Appreciating the principles of chaos theory also fosters an appreciation for the complexity and interconnectedness of our world. It brings to light the delicate balance within ecosystems and the potential impact of small changes, underscoring the importance of sustainability and environmental stewardship.

Ultimately, engaging with chaos theory encourages innovative thinking and problem-solving techniques that are non-linear and non-traditional. This novel approach is crucial for tackling modern-day challenges in a comprehensive and holistic manner, ensuring that as new complexities arise, we are better equipped to understand and manage them effectively.

As humans continue to progress and build increasingly complex systems, the lessons from chaos theory will only grow in relevance, making its study not just beneficial but essential for future advancements.

This section integrates seamlessly into the progression of the book's chapter on chaos theory by detailing the practical implications and necessity of understanding this intricate field, preparing readers for subsequent discussions on applied chaos theory in various domains.

Chapter 2

The Butterfly Effect: Sensitive Dependence on Initial Conditions

This chapter delves into one of the most compelling aspects of chaos theory, which highlights how small changes in initial conditions can lead to vastly different outcomes, a phenomenon often encapsulated by the metaphor of a butterfly's wings causing hurricanes. It traces the concept's origins, its mathematical formulation, and its pervasive influence across various disciplines, demonstrating how this principle shapes our understanding of complex systems and unpredictability.

2.1 Defining the Butterfly Effect

The Butterfly Effect refers to a fundamental concept in chaos theory, which articulates that small changes in the initial conditions of a dynamic system can result in large and unforeseen variations in later states. The name of the effect, poetic in its imagery, is derived from the metaphorical example that the flapping of a butterfly's wings might ultimately cause a hurricane on the other side of the world. This vivid illustration underscores the interconnectedness and sensitive dependencies in chaotic systems.

In a more formalized context, the Butterfly Effect is associated with systems that exhibit deterministic nonlinearity, meaning they follow specific rules but can evolve in seemingly random and unpredictable ways due to their inherent sensitivity to initial conditions. This sensitivity indicates that even minuscule variations in the starting point of a system can lead to exponentially divergent outcomes, rendering long-term prediction nearly impossible beyond a certain temporal horizon.

To consider this concept mathematically, let us envisage a dynamic system described by a set of differential equations. The trajectory of the system in its phase space (a multidimensional space in which all possible states of the system are represented) can be drastically altered by initial conditions—numerically represented by even the smallest decimal places.

One can visualize this by plotting the trajectories of two nearly identical initial points on a graph and observing how, over time, these trajectories diverge. In systems characterized by the Butterfly Effect, these paths do not just separate; they do so exponentially, reflecting the system's sensitive dependence. For visualization, employing a graphical tool such as TikZ is ideal:

2.1. DEFINING THE BUTTERFLY EFFECT

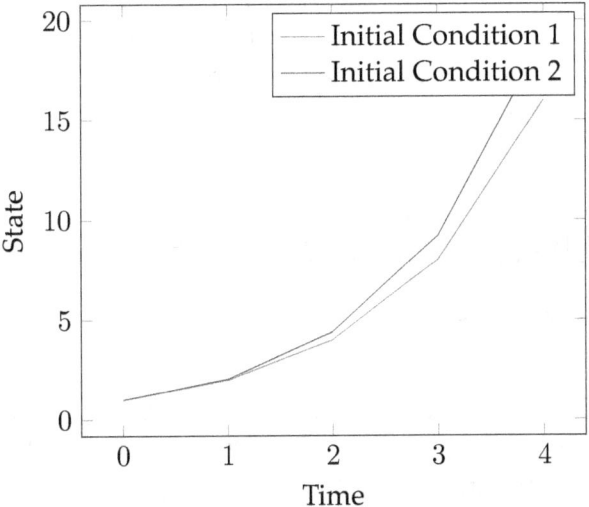

Trajectory of a Dynamical System Exhibiting the Butterfly Effect

This diagram highlights how small differences in initial conditions (red and blue lines starting close to each other) diverge rapidly over time, illustrating the chaotic nature of such systems.

The implications of the Butterfly Effect are profound because they challenge the predictability and deterministic understanding of complex systems. In fields ranging from meteorology to economics, recognizing the limitations imposed by this effect is crucial for developing more robust predictive models that can accommodate and account for inherent unpredictabilities.

As we continue to explore the nuances of the Butterfly Effect, it becomes increasingly clear that this phenomenon is not just about theoretical mathematics or abstract models but about a fundamental characteristic of the natural world. It challenges our perceptions of causality and control, suggesting that the universe might be more interconnected and unpredictable than previously assumed. This perspective invites further investigation into how we understand and interact with complex systems, emphasizing a cautious approach towards prediction and manipulation based on initial conditions.

2.2 Origins and Development of the Concept

The origin of the Butterfly Effect is inextricably linked to the development of chaos theory itself, rooted in the work of mathematician and meteorologist Edward Lorenz. In the early 1960s, Lorenz was exploring numerical weather prediction at the Massachusetts Institute of Technology. His research inadvertently led to the discovery of deterministic chaos, a groundbreaking insight into how nonlinear systems behave. Lorenz's seminal moment occurred when he attempted to re-run a weather simulation using what he believed were identical initial conditions. However, due to a slight rounding in the numerical precision of the data (from six decimal places to three), the outcome of the model diverged spectacularly from the original run. This led him to hypothesize that small changes in initial conditions could yield profoundly different outcomes, which could not be predicted by linear extrapolation from previous states.

As Lorenz delved deeper, he formulated the conceptual foundations of the Butterfly Effect, famously presenting a talk titled "Predictability: Does the Flap of a Butterfly's Wings in Brazil Set Off a Tornado in Texas?" at the 139th meeting of the American Association for the Advancement of Science in 1972. This talk not only popularized

2.2. ORIGINS AND DEVELOPMENT OF THE CONCEPT

the Butterfly Effect but also marked a pivotal moment in the broader acceptance and understanding of chaos theory.

The mathematical underpinning of this concept, later detailed in Lorenz's 1963 paper "Deterministic Nonperiodic Flow," was equally revolutionary. Here, Lorenz presented his findings about the Lorenz Attractor, a set of chaotic solutions for a simplified model of atmospheric convection. The attractor illustrated how small differences in initial trajectories could diverge exponentially over time, effectively visualizing the essence of the Butterfly Effect in mathematical terms.

Lorenz's discovery was fortified by contemporaries across various scientific fields who recognized similar patterns beyond meteorology. Biologists, economists, and engineers began to note unpredictable behaviors in their own systems, which seemed governed by sensitivity to initial conditions, thereby expanding the scope and relevance of Lorenz's work. This interdisciplinary cross-pollination further developed through the 1970s and 1980s, as chaos theory and the Butterfly Effect moved beyond theoretical exploration and into practical applications. Through conferences, scholarly articles, and a heightened interest in nonlinear dynamics, knowledge of the concept expanded. Innovations in computational power during this era also allowed for more complex models and simulations that could more accurately represent chaotic systems and investigate the implications of small initial differences.

Statisticians and mathematicians like Benoit Mandelbrot also pushed the boundaries of chaos and fractals, introducing geometric dimensions to the discussions around chaotic behavior and further embedding the principles of the Butterfly Effect in scientific thought. Mandelbrot's work on fractals provided a new lens through which to view the erratic yet patterned nature of chaotic systems, helping to illustrate why such systems might be sensitive to their starting points.

These developments collectively elaborated on Lorenz's original insights, propelling the Butterfly Effect from a theoretical curiosity to a cornerstone concept in understanding complex systems across diverse disciplines. Exploring this concept has not only shed light on the unpredictability inherent in many natural and human-made systems but has also challenged previous deterministic predictions about the behavior of these systems over time.

2.3 Mathematical Foundations of the Butterfly Effect

The mathematical grounding of the Butterfly Effect finds its roots in the study of dynamical systems, particularly through differential equations that describe how a system evolves over time. Central to understanding this phenomenon is the concept of sensitivity to initial conditions, a fundamental characteristic of chaotic systems.

To dissect this concept mathematically, let us consider a dynamical system modeled by a set of deterministic differential equations:

$$\frac{dx}{dt} = f(x(t), t)$$

where $x(t)$ represents the state of the system at time t, and f is a function that describes the system's evolution laws.

2.3. MATHEMATICAL FOUNDATIONS OF THE BUTTERFLY EFFECT

For simplicity, assume x and t are real numbers, although in complex systems, they could be vectors and functions could be multidimensional.

Chaos manifests in such systems when small changes in the initial state $x(0)$ lead to exponentially diverging outcomes. This divergence can be quantitatively defined through what is known as Lyapunov exponents, which measure the rate at which nearby trajectories diverge. For a chaotic system, at least one positive Lyapunov exponent is typically present. This exponent, λ, is defined as:

$$\lambda = \lim_{t \to \infty} \frac{1}{t} \log \left(\frac{|\delta x(t)|}{|\delta x(0)|} \right)$$

where $\delta x(0)$ and $\delta x(t)$ represent the initial and evolved differences in the states of two trajectories that started extremely close to each other.

To visualize the impact of this sensitive dependence, consider the famous logistic map, an archetype example of how simple nonlinear dynamical systems can exhibit chaotic behavior. The map is defined as:

$$x_{n+1} = r x_n (1 - x_n)$$

Here r is a parameter that affects the system's behavior, and x_n is confined within the interval $[0,1]$. As r increases, the system transitions from stable and periodic behaviors to chaotic dynamics. Notably, when r is around 3.57, the system begins to display chaotic characteristics where tiny differences in initial values x_0 dramatically influence the trajectory, showcasing the Butterfly Effect.

In computational studies, determining the predictability of such systems involves simulating trajectories with slightly varied initial conditions. Many software tools and systems built for chaos analysis allow researchers to

CHAPTER 2. THE BUTTERFLY EFFECT: SENSITIVE DEPENDENCE ON INITIAL CONDITIONS

plot these trajectories and observe how they evolve differently over time. Below is an example using Python's matplotlib and numpy libraries to demonstrate this:

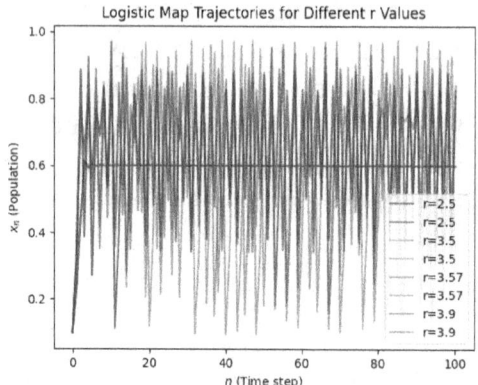

Here, each value of r generates a distinctly different pattern, emphasizing how initial conditions and system parameters crucially influence outcomes. Through such examples and computational exploration, one appreciates the depth and implications of the Butterfly Effect in mathematical terms, precisely capturing the essence of sensitive dependence and unpredictability in chaotic systems.

Through detailed mathematical explanations and examples, this section elucidates the rigorous and intriguing underpinnings of the Butterfly Effect, highlighting its role as a cornerstone concept in the field of chaos theory.

2.4 Examples in Meteorology

Meteorology, the science of the atmosphere, was one of the first fields to provide a real-world context to the theoretical implications of the butterfly effect. This con-

2.4. EXAMPLES IN METEOROLOGY

nection is not just theoretical; it has practical implications in weather forecasting and understanding severe weather phenomena. We delve into several pivotal studies and real-life examples that highlight how meteorology both demonstrates and leverages the butterfly effect in its methodologies.

1. Forecasting Weather: The most straightforward example of the butterfly effect in meteorology is weather forecasting. Early weather models struggled with accuracy beyond a few days. Edward Lorenz, a meteorologist and mathematician, discovered through his work in the early 1960s that tiny rounding errors in his computational models could result in vastly different weather outcomes. This revelation stemmed from Lorenz's observation of simulations where a slight change — as minimal as a change from .506127 to .506 — drastically altered the forecast. This sensitivity to initial conditions highlighted that long-term weather predictions could be inherently unpredictable beyond a certain temporal threshold due to an accumulation of minute and initially insignificant variations.

2. Hurricane Path Prediction: The path of hurricanes can be significantly influenced by infinitesimal environmental variations. Initially, tiny fluctuations in air temperature, water vapor levels, or even minor changes in ocean currents can dramatically alter the trajectory and intensity of cyclonic systems. For example, consider the forecasting challenges of Hurricane Katrina in 2005. Weather models initially had difficulty agreeing on the storm's path due to slight differences in initial data inputs. As we observed, these initial discrepancies had profound impacts on predictions of landfall location, thus affecting evacuation orders and emergency responses.

3. Tornado Formation: Tornado genesis is another area where the butterfly effect is strikingly evident. Slight variations in atmospheric conditions, such as temperature gradients and wind shear, can determine whether a tornado forms, its path, and its longevity. Researchers utilizing Doppler radar and other sensor data have noticed that very small differences in initial conditions can either favor or hinder the formation of these destructive wind funnels. Such findings emphasize the chaotic nature of tornado formation and underscore the broader challenges of severe weather prediction.

To illustrate these concepts visually, let us examine a simple simulation using data from a published meteorological study. The data represent two scenarios with marginally different initial conditions:

Scenario	Initial Condition Variation	Outcome
1	Temperature gradient of 0.1°C	Development of a tornado
2	Temperature gradient of 0.2°C	No tornado development

This table represents a simplified version of real-life data collection scenarios where seemingly negligible differences lead to distinct outcomes. In comprehensive weather models, hundreds such factors play out in complex ways, showcasing sensitivity to initial conditions at a grand scale.

As we progress further in understanding and refining meteorological models through technologies like AI and massive data processing, we acknowledge the butterfly effect's role as both a complicator and a critical tool, guiding better predictions and smarter responses to weather-related events. By recognizing the embedded chaos in atmospheric behaviors, meteorologists continue to refine techniques that may one day enhance forecast reliability despite inherent unpredictabilities.

2.5 Experimental Observations and Evidence

The practical observations and empirical evidence underpinning the Butterfly Effect offer fascinating insights into the inherent unpredictability of complex systems. While theoretical work provides the framework, it is through experimental observations that the depth of sensitivity to initial conditions is truly uncovered.

One of the most notable early experiments that illustrated this concept was conducted by meteorologist Edward Lorenz in the 1960s. Using a simple computer model to simulate weather patterns, Lorenz discovered that minuscule changes in input conditions could diverge wildly in output, leading to drastically different weather scenarios. This experiment used a deterministic, non-linear system consisting of twelve equations to simulate air movement and was intended to predict weather by accurately inputting initial conditions. However, Lorenz noted that even rounding variables to a slightly different number of decimal places led to significantly different weather outcomes. This served as one of the first empirical confirmations that small variances in initial conditions could yield large-scale divergence in complex systems.

Following Lorenz's pioneering work, numerous experiments across various scientific domains have demonstrated similar results, reinforcing the universality of the Butterfly Effect. In physics, for instance, experiments with double pendulums or chaotic water wheels show how initial positioning can affect the dynamic evolution of the system, leading to unpredictable motion patterns.

In ecology, researchers have exposed how initial population densities can influence long-term dynamics of ecosystems, potentially leading to different evolutionary outcomes. A slight increase in the population of a particular species, under certain conditions, might result in an uncontrolled population explosion or lead to its unexpected extinction due to the intricate balance of ecological networks.

Moreover, the field of electronics provides illustrative examples through circuit experiments. Chaotic circuits, or those that exhibit sensitivity to initial conditions, can alter their voltage or current outputs dramatically based on minute changes at the start. These experiments not only corroborate Lorenz's findings but also expand them by illustrating how the Butterfly Effect can manifest in non-meteorological systems.

The medical field is not exempt from this phenomenon either. Recent studies in epidemiology show that slight variations in patient zero's condition or timing of infection can drastically change the spread patterns of infectious diseases, impacting outbreak containment strategies.

These diverse experiments share a common theme: they underscore the challenge of prediction in complex systems where initial conditions cannot be determined with absolute precision. The implications are profound, affecting everything from weather forecasting to understanding economic fluctuations and managing pandemics.

To visually represent this concept, consider the following simple system modeled using a set of differential equations to exhibit chaotic behavior:

$$\dot{x} = \sigma(y-x), \quad \dot{y} = rx - y - xz, \quad \dot{z} = xy - bz$$

These are the Lorenz equations, a simplified model for atmospheric convection where σ, r, and b are parameters indicating the system's sensitivity to initial conditions. Plotted over time, the solutions to these equations depict what is known as the Lorenz attractor, a set of chaotic solutions that demonstrate how paths diverge over time from closely aligned initial points.

This robust body of experimental evidence across numerous fields not only validates the concept of the Butterfly Effect but also illuminates the intricate and unpredictable nature of our world. It emphasizes the necessity of advancing our methods of modeling complex systems to better accommodate and predict the cascading effects triggered by minuscule variations at any given point in time. By understanding and anticipating such dynamics, perhaps we can navigate more adeptly through the inherent unpredictabilities of natural and human-made systems.

2.6 Butterfly Effect in Other Systems

The Butterfly Effect, while profoundly rooted in the realm of meteorology and mathematical chaos theory, extends its reach far beyond these borders. This influence can be traced in various systems across different disciplines, including economics, biology, and engineering, demonstrating the universal applicability of sensitivity to initial conditions.

In the economic sphere, the Butterfly Effect illuminates the intricate and often precarious nature of financial markets. Consider the stock market: a rumor or even a minor policy change by the Federal Reserve can result in massive fluctuations. For example, in 2015, the Swiss National Bank's unexpected decision to decouple from the Euro led to dramatic, swift, and far-reaching impacts on global currency markets. Here, we see an analogous pattern to weather systems—a small trigger can lead to vast economic storms.

Biology offers another fascinating canvas showcasing this effect. Ecosystems are highly complex and sensitive systems where small alterations can lead to radical shifts in population dynamics and biodiversity. A notable illustration is the introduction of a non-native species into a region, which might initially seem insignificant but can disrupt local ecosystems profoundly, often leading to unexpected and severe consequences for species abundance and health. This phenomenon echoes the chaotic ripple effects observed in weather patterns triggered by minute initial changes.

Furthermore, the engineering field is replete with examples where initial conditions play a critical role in system behavior, especially in large-scale structures and computer networks. Small defects in material structure can significantly affect the integrity and failure modes of engineering constructions. The collapse of the Silver Bridge in West Virginia in 1967 serves as a stark reminder. A minor flaw in a single eyebar in a suspension chain set off a catastrophic failure of the bridge, demonstrating how minute initial structural imperfections can provoke large-scale disasters.

To illustrate the interconnected and unpredictable nature

of systems influenced by the Butterfly Effect, consider employing a Lorenz attractor diagram"

Simplified 3D Plot

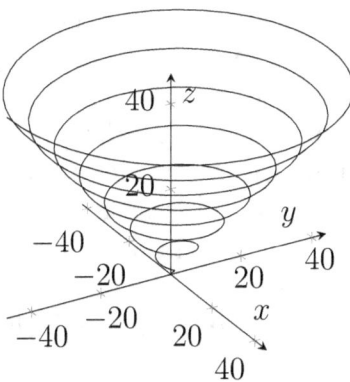

Each of these examples underscores that our world is rife with systems that, though they may appear stable and predictable at a glance, are fundamentally influenced by initial conditions in profoundly unpredictable ways. Spanning across various fields, these instances reveal the underlying complexity and interconnectedness of different systems within our ever-changing world.

In rightfully appreciating the scope of the Butterfly Effect across these divergent systems, we gain insights not only into their individual complexities but also into the potential for seemingly trivial factors to engender significant consequences. This understanding fosters a broader comprehension of the dynamics at play, which can aid in better management and anticipation of the effects in complex systems. Through this lens, we see the world as an intricately woven tapestry of cause and effect—a fabric that is continuously shaped by the flutters of countless metaphorical butterflies.

2.7 Implications for Predictive Models

The Butterfly Effect, emphasizing the sensitivity to initial conditions, has profound implications for predictive models in various scientific and engineering fields. In this discussion, we turn our attention to how this concept challenges and refines the accuracy of models designed to forecast future states in systems ranging from weather predictions to stock market forecasts. This scrutiny of model predictability not only elucidates the limitations inherent in these approaches but also inspires the advancement of new modeling techniques that embrace aspects of chaos theory.

For decades, meteorologists have strived to predict weather patterns with precision and accuracy. Traditionally, these models involved inputting current weather data into a set of equations that simulated atmospheric physics. However, the inception of chaos theory, particularly the Butterfly Effect, demonstrated that infinitesimally small measurement errors—no larger than the impact of a butterfly flapping its wings—could drastically alter the predicted weather outcome days or weeks later. This sensitivity poses a significant challenge: how can we predict weather accurately when such minute discrepancies can escalate into substantial impacts?

In response, meteorologists have begun to employ ensemble forecasting systems. These systems use multiple simulations (ensembles) that start from slightly varied initial conditions typically reflective of the inherent uncertainties. Each different version of the initial condition can lead to divergent forecast paths, providing a spread of possible outcomes. Presenting these as probability distri-

2.7. IMPLICATIONS FOR PREDICTIVE MODELS

butions rather than single-point forecasts, meteorologists can offer a more realistic range of potential weather developments. This probabilistic approach does not defeat the uncertainty caused by the Butterfly Effect but rather incorporates it into the forecast, offering a clearer picture of possible futures.

In financial markets, predictive models face a similar conundrum. Stock market forecasting models also suffer from an acute sensitivity to initial conditions. Small events or transactions might trigger significant economic repercussions, a scenario vividly illustrated by the rapid stock market crashes prompted by seemingly minor causes. Here, traditional mathematical models that predict based on historical data and trends find themselves inadequate when a tiny, unpredictable event can pivot the market direction dramatically.

Finance professionals have thus been nudged towards adopting more robust statistical methods and machine learning techniques that account for large volumes of data and recognize patterns that suggest potential market changes. These methods aim to absorb and adapt to the volatility and apparent randomness inherent in financial systems influenced by the Butterfly Effect.

Moreover, the concept has reshaped the approach to predictive modeling in ecology, where understanding species population dynamics or ecosystem changes with precision is critical. Ecological models now often incorporate stochastic elements—randomly determined inputs that reflect natural variability and chaotic behavior. This approach acknowledges that small changes, such as a minor temperature fluctuation or a single species' sudden population decline, can lead to dramatically different ecological states.

Emerging from these discussions is an evident shift in predictive modeling across various fields, a shift from a deterministic to a probabilistic or stochastic framework. This transition acknowledges that absolute certainty in predictions is unachievable and embraces the complexity and inherent unpredictability of natural and human-made systems. By integrating chaos theory principles into predictive models, scientists and professionals can develop more resilient strategies that accommodate and make use of the unavoidable uncertainty highlighted by the Butterfly Effect.

The exploration of these revised predictive strategies reveals an essential truth about our interaction with complex systems: while we cannot control all variables or predict all outcomes, we can improve our understanding and management of these systems by acknowledging and planning for the unpredictable. This shift not only enhances model reliability across diverse disciplines but also deepens our conceptual grasp of how chaos molds the natural and abstract landscapes we navigate daily.

2.8 The Butterfly Effect in Popular Culture

The Butterfly Effect, a term popularized in chaos theory but familiar far beyond scientific circles, has permeated various facets of popular culture, influencing literature, film, music, and public discourse. Its profound concept, that small causes can have large effects, sparks not only scientific curiosity but also resonates deeply with the human experience of unforeseeable consequences.

Firstly, the impact of the Butterfly Effect is notably prominent in the world of cinema. The 2004 film titled "The Butterfly Effect," starring Ashton Kutcher, explicitly draws on this concept. In the movie, the protagonist discovers his ability to travel back in time and alter past events, only to find that each change has far-reaching, often devastating consequences in his present life. While the scientific underpinnings in the film are loosely interpreted, the central premise underscores a dramatic interpretation of chaos theory's sensitivity to initial conditions. This movie, and its subsequent sequels, vastly amplified the layperson's awareness and understanding of the Butterfly Effect, turning a complex scientific principle into a household term associated with unpredictability and dramatic shifts resulting from seemingly minor actions.

In literature, authors have long employed elements of the Butterfly Effect, integrating its core message into narratives that explore the complex web of human decisions and fate. A salient example is Ray Bradbury's short story "A Sound of Thunder." The story features time travelers who visit the prehistoric past to hunt a Tyrannosaurus rex, only to find out that a minor mishap - the accidental crushing of a butterfly - alters the course of history dramatically. Bradbury's articulation effectively communicates the delicate balance of ecosystems and historical trajectories, making a poignant comment on the importance of each individual action within a broader temporal landscape.

Moreover, the concept has been a recurrent theme in discussions about ecological awareness and environmental advocacy. The narrative that a small act, like deciding to recycle a piece of plastic, could potentially lead to significant positive outcomes for planetary health incorporates the principles of the Butterfly Effect. This use in environmental campaigns serves to empower people, suggesting that their small, seemingly insignificant actions can collectively lead to significant environmental benefits.

Music and other forms of artistic expression have also embraced this theme. The notion that small initial differences can lead to overwhelming consequences provides rich metaphorical content for lyrics and music videos, often reflecting on personal life choices and their unforeseen impacts on personal or collective futures.

The pervasive appeal of the Butterfly Effect in popular culture is perhaps due to its dual allure: it captures a fundamental truth about the natural world while also engaging with deep existential queries about control, fate, and consequence in human lives. It underscores a fascinating intersection between deterministic laws and the unpredictable flourishes that seem to characterize much of human experience.

As this phenomenon continues to echo through various cultural expressions, it not only reinforces the relevance of chaos theory but also enriches our discourse about the impact of human actions and the unpredictability inherent in our complex world.

Chapter 3

Fractals: The Patterns of Chaos

This chapter explores fractals, intricate patterns that repeat at every scale and are considered a key feature of chaos. It covers the fundamental properties and types of fractals, the methods used for generating them, and their practical implications. The discussion extends to how fractals manifest in nature and the various applications in technology, art, and scientific research, providing a comprehensive understanding of their role in revealing the inherent order within chaotic systems.

3.1 Introduction to Fractals

Fractals are intricate structures that emerge out of simple repetitive processes, typically characterized by their self-similar patterns irrespective of the scale of observation. Derived from the Latin word *fractus* meaning "broken" or "fractured", fractals were propelled into mainstream scientific research by Benoit Mandelbrot, who famously defined them as "a rough or fragmented geometric shape that can be split into parts, each of which is (at least approximately) a reduced-size copy of the whole." This foundational concept opens up a mesmerizing universe where the rules of traditional geometry are stretched and reevaluated.

One of the fundamental characteristics of fractal geometry is its defiance of classical dimensionality. Unlike traditional geometric figures like lines, squares, or cubes, which have integer dimensions of 1, 2, and 3 respectively, fractals exhibit non-integer dimensions. The concept of fractal dimensions introduces a quantitative measure that describes how completely a fractal appears to fill space, as one zooms down to finer and finer scales. This intricate expansion is defined through a logarithmic relationship between the feature size and the number repeated. Mandelbrot's pioneering work not only introduced the term but also provided a mathematical form to describe these anomalous dimensions, which is crucial in quantifying the complexity inherent in fractals.

While fractals are mathematically fascinating and complex, their generation can be surprisingly simple. This accessibility is achieved through recursive algorithms that iterate over a basic process; each iteration leads to greater complexity. The process often begins with a straightforward rule or equation, such as the iteration $z_{n+1} = z_n^2 + c$ used in generating the Mandelbrot set. Each repetition of the rule adds a layer of complexity, illustrating a key property of fractals: intricate behaviors emerging from straightforward rules.

Moreover, in visualization, even though individual fractals may differ vastly in appearance—ranging from the infinitely detailed edges of the Mandelbrot set to the delicate, fern-like patterns of the Barnsley Fern—there exists a typical aesthetic of rugged, chaotic formations that strangely resonate a sense of harmony and balance. This is because the elemental patterns continually repeat, regardless of the scale at which they are viewed.

In considering fractals within the broader scope of scientific inquiry, they serve as a bridge linking seemingly unrelated phenomena across nature and various scientific disciplines. The intrinsic properties of fractals allow them to model complex structures and patterns in nature such as coastlines, mountain ranges, and even patterns of galactic clustering, providing insights into the scaling laws that govern these phenomena.

Moreover, fractals have transcended their geometric roots to permeate various fields such as algorithmic art, digital imaging, signal processing, and more. They provide a framework where the simplicity and complexity coexist, driven by recursion, scale independence, and self-similarity — foundational concepts that encourage a deeper exploration of the chaotic yet orderly universe we inhabit.

In delving deeper into this section, we have laid the groundwork necessary to understand more complex manifestations and applications of fractals, which will be explored in subsequent sections of this chapter.

3.2 Properties of Fractals

Fractals are fundamentally distinguished by characteristics that set them apart from traditional geometric figures. One of the most notable properties of fractals is their **self-**

similarity. Self-similarity in fractals may be exact or statistical. In exact self-similarity, each part of the fractal is an exact smaller copy of the whole, common in mathematical fractals like the Sierpiński triangle. In natural fractals, however, this property often manifests as *approximate* self-similarity where the resemblance is statistical and not necessarily precise, particularly over different scales.

Another key property of fractals is their **infinite complexity**. This term refers to the fact that fractals can exhibit details at arbitrarily small scales. No matter how much you zoom in, new intricate patterns emerge, each layer revealing more complexity. This inherent complexity is quantified through what is known as the **fractal dimension**, which intriguingly can be a non-integer. This dimension is a ratio providing a statistical index of complexity, comparing how the detail in a fractal changes with the scale at which it is measured.

Fractal dimensions are not typically whole numbers. The dimension of a line is 1, a plane is 2, but fractals usually have dimensions that fall somewhere in between, such as 1.58 or 2.43. There are various methods to calculate the fractal dimension, such as the box-counting dimension where one covers the fractal with boxes of a certain size and counts how many boxes contain a part of the fractal. The math behind this can be rather elaborate, but essential for understanding the true nature of fractals.

Fractals also display a property referred to as **scale-invariance**, stemming from their self-similarity. This means patterns look similar at various magnifications or reductions. Such behavior can not only be visually mesmerizing but also mathematically significant as it hints at underlying invariables despite the apparent randomness or chaotic nature at initial glances.

Fractals' **emergence** property illustrates their capacity to emerge through simple iterative processes yet resulting in complex final outcomes. This emergence is often referred to when discussing natural processes and explains how intricate structures such as coastlines, snowflakes, and even cardiovascular and pulmonary structures form.

Understanding these properties not only unveils the mystical allure of fractals but also lays a foundation for appreciating their practical applications in science, technology, and art. By studying the recursive nature and dimensional intricacies of fractals, important insights can be gained into both the construct of synthetic systems and natural phenomena. Thus, fractals bridge the conceptual with the tangible, offering a fascinating glimpse into the complexities that govern various branches of science and influencing technological advancements in rendering detailed and effective solutions in fields as diverse as graphics design and environmental science.

3.3 Generating Fractals: Methods and Techniques

Delving deeper into the realm of fractals requires not only an understanding of their theoretical basis but also practical methods for their generation. Fractal generation techniques are as varied as the fractals themselves, emphasizing the symbiotic relationship between simplicity and complexity characteristic of fractal geometry. These methods range from recursive algorithms to iterative systems, each manifesting a distinct type of fractal geometry.

Iterated Function Systems (IFS): The cornerstone technique for fractal generation is the Iterated Function System (IFS), which uses fixed geometric replacement rules. IFS comprises multiple contraction mappings on a complete metric space, which, when applied repeatedly, generate a fractal. A classical example is the Sierpinski Triangle, where a triangle is recursively subdivided into smaller triangles. The mathematical elegance of IFS lies in its basis on simple transformations: translation, rotation, scaling, and shearing, which collectively enable the creation of highly complex structures.

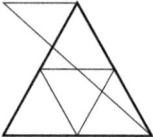

Figure 3.1: First iteration in generating a Sierpinski Triangle using IFS.

Escape-time Algorithms: Popular in generating famous fractals like the Mandelbrot and Julia sets, this method involves iteratively applying a function to each point in a complex plane and recording the time it takes to escape a certain condition. The result is a stunning visualization of boundary dynamics as points on the edge of escape paths exhibit intricate, never-ending patterns.

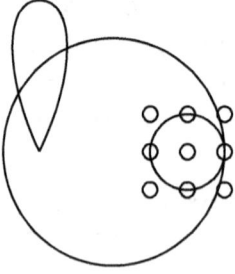

Figure 3.2: Stylized representation of the Mandelbrot set.

3.3. GENERATING FRACTALS: METHODS AND TECHNIQUES

Random Fractal Generation: Not all fractals are deterministic. Random fractal generation uses probabilistic rules to modify each iteration's output slightly or dramatically, leading to a variety of realizations under the same general guidelines. Techniques such as the random midpoint displacement algorithm are utilized in generating landscapes and natural phenomena in virtual environments. This method involves displacing midpoints of lines or surfaces randomly at each iteration, thereby creating realistic terrain models.

L-systems or Lindenmayer Systems: Originally developed to describe the behavior of plant cells, L-systems provide a powerful tool to model the growth processes of plants using string rewriting. L-systems use an iterative approach to replace parts of an initial simple string with more complex sequences, leading to intricate patterns reminiscent of natural growth processes.

The ingenuity of fractals lies not merely in their aesthetic or theoretical appeal but in the myriad ways they can be realized, each method presenting a unique perspective and insight into the underlying order of chaos. As mathematicians and artists alike delve into these generative pathways, they reveal more than just shapes and patterns; they unravel an entire universe governed by recursive harmony and complexity.

Let us remember that every method described embodies the essential enigma of fractals: simple rules giving rise to endlessly complex forms—a powerful analogy for countless phenomena in our universe.

3.4 Famous Fractals and Their Characteristics

Fractals, with their self-similar patterns, are not only an academic interest but also an aesthetic spectacle that has captivated the attention of both scientists and artists. Among the multitude of fractals, a few stand out prominently due to their distinct characteristics and historical significance. We explore some of these famous fractals, their intrinsic properties, and the equations that govern their formation.

The Mandelbrot Set: Named after Benoit Mandelbrot, who visualized and defined this complex structure, the Mandelbrot set is arguably the most recognized fractal. It is defined in the complex plane, where a certain set of complex numbers, c, form a fractal when the sequence $z_{n+1} = z_n^2 + c$ remains bounded, meaning it does not diverge to infinity. The boundary of the Mandelbrot set is exquisitely complex and infinitely detailed. This fractal acts as a catalog for Julia sets; if you zoom into the boundary of the Mandelbrot set, you can see various forms of corresponding Julia sets unfold.

Julia Sets: Created by Gaston Julia, these fractals are closely related to the Mandelbrot set but are generated by fixed points from a complex quadratic polynomial. The general form considered is $z_{n+1} = z_n^2 + c$, where c is a constant complex number and z varies over each point in the plane. The nature of the Julia set's boundary depends on the choice of c, which can produce dramatically different forms—ranging from highly connected figures to completely disconnected dust-like structures.

The Sierpinski Triangle: This fractal is a prime example

3.4. FAMOUS FRACTALS AND THEIR CHARACTERISTICS

of a self-similar set, simple yet profound. It starts with an equilateral triangle, and the fractal pattern is produced by recursively removing the upside-down triangles from the center, leaving three adjoining triangles in the first iteration. This process is repeated indefinitely. The Sierpinski Triangle can also be described through its Hausdorff dimension, which is approximately 1.585, providing insight into the scaling properties at different iterations.

The Koch Snowflake: First described by Helge von Koch, this fractal begins with a simple equilateral triangle. Each iteration involves adding a smaller equilateral triangle to the middle third of each side of the figure, leading to a snowflake-like pattern with an infinitely jagged perimeter. Intriguingly, while the area of the Koch snowflake remains finite, its perimeter grows indefinitely with each iteration.

The Cantor Set: Discovered by Georg Cantor, this fractal is subtle and profound, consisting of an interval that gets iteratively bisected with the central third removed. Although its construction process is simple—keep removing the middle third of each segment—the Cantor set introduces complex concepts such as those involving measure theory and topology, including its property of being uncountably infinite yet having zero measure.

These fractals are more than just curiosities; they are also fundamentally essential in analyzing and understanding the limits and extensions of mathematical and geometrical concepts. Each fractal discussed provides a unique insight into the complexity that can emerge from surprisingly simple iterative processes. As we scale down or up within these self-replicating structures, new mysteries and beauty unfold at every turn, reflecting the intrinsic unpredictability and order within systems governed by chaos theory.

3.5 Fractals in Nature

Fractals, with their infinitely complex patterns, are not just a mathematical curiosity but a fundamental facet of the world around us. Nature, often irrational and wild in its arrangement, adheres to patterns and scales that manifest fractals in a multitude of forms. These patterns are not only aesthetically breathtaking but also offer a canvas where the theories of chaos and order merge.

One of the most easily recognizable examples of fractals in nature is the branching pattern of trees and plants. The recursive process of branch splitting in trees is a classic case where a single simple rule results in a complex pattern. Each bifurcation or splitting leads to two or more smaller branches, and this pattern repeats over and over on different scales. This self-similarity is one of the core principles of fractal geometry. One might explore the L-system, a mathematical system used to model the growth processes of plants, to better understand this phenomenon. The L-system uses an initiator and generator mechanism which effectively encodes the recursive nature of tree growth with great accuracy.

Another quintessential example of natural fractals is found in the structure of Romanesco broccoli, which presents an almost perfect fractal pattern. Each bud is composed of a series of smaller buds, arranged in a logarithmic spiral. This pattern repeats itself in such a way that the visual complexity of the Romanesco is made manifest at every scale, from the whole vegetable down to the smallest bud.

3.5. FRACTALS IN NATURE

Coastlines represent another profound instance of fractal geometry in nature; their calculated fractal dimensions can vary, illustrating that the finer the measurement scale, the longer the coastline appears—reminiscent of the "Coastline paradox" described by mathematician Benoit Mandelbrot. Mandelbrot's studies showed that measuring a coastline with a smaller ruler results in a longer perceived length, highlighting how fractal geometry can be a tool to reconcile the seemingly unmeasurable.

Moving from earth-bound instances to the cosmic scale, fractal patterns are also observable in the distribution of matter in the universe. Galaxies and large-scale structures exhibit clustering patterns that hint at fractal-like processes in play. Researchers have used fractal dimensions to study the structure of the universe itself, exploring how mass is distributed across different scales.

On a smaller scale, snowflakes are among nature's most delicate fractal patterns. Each snowflake is made up of ice crystals that form around dust particles in the atmosphere. The crystalline water molecules adhere to hexagonal symmetry due to hydrogen bonds, but as they grow, the interplay of temperature and humidity forces each snowflake to adopt unique patterns that are iteratively repeated.

At the intersection of function and design, fractals also appear in the animal kingdom. The patterns on certain shells, like that of the nautilus, are governed by logarithmic spirals that reflect fractal properties. Additionally, the way blood vessels and bronchial tubes branch throughout the body mirrors fractal geometry aimed at maximizing surface area in a limited space, vital for efficient circulation and air distribution.

What becomes evident through these examples is that

fractals are not merely repetitive patterns, but are a necessary form of complexity that nature employs to increase efficiency, optimize space, and possibly even create beauty. The self-replicating precision that seems random yet ordered reveals much about how natural phenomena are structured—a testament to the underlying rules driving the chaos we often witness in natural environments.

This deep interconnection between fractals and natural phenomena not only enhances our understanding of nature's design but also propels scientific discovery, allowing us to see order in what seems initially to be disarray. The implications of this reach far beyond aesthetic appreciation, touching on critical scientific inquiries and enhancing practical applications, from designing more efficient transport networks to understanding the spread of diseases in populations.

3.6 Applications of Fractals in Science and Technology

Fractals, with their self-similar patterns and intrinsic complexity, have found a variety of profound applications in many fields of science and technology. Their unique properties allow them to model complex structures and behaviors in a more natural and efficient way than traditional geometric shapes.

Medicine and Biology: In medicine and biology, fractals provide significant insights due to their appearance in natural organisms and physiological processes. For example, the branching patterns of lungs and blood vessels, which are crucial for maximizing the exchange surfaces

3.6. APPLICATIONS OF FRACTALS IN SCIENCE AND TECHNOLOGY

without occupying large volumes, can be effectively described using fractal geometry. This description not only helps in understanding the growth patterns but also aids in diagnosing diseases that cause structural changes in these organs. Additionally, the application of fractal analysis in oncology, to study the irregular growth patterns of tumors, has proven crucial in identifying malignant changes at early stages.

Telecommunications: Fractal shapes are used to design more efficient and compact antennas. Due to their self-similar nature, fractal antennas can operate at multiple frequencies and are therefore excellent for cellular phones, radios, and broadband internet devices. This multiband capability, coupled with a smaller size compared to traditional antenna designs, makes fractal antennas highly desirable in communications where space and bandwidth are at a premium.

Material Science: In materials science, researchers leverage fractal concepts to create more resilient and adaptable materials. The fractal design principles have been applied to develop lightweight structural composites that exhibit enhanced strength and flexibility for aerospace and automotive applications. Furthermore, the thermal properties of materials can also be tuned by manipulating their fractal dimensions, which can lead to new technologies for thermal management in electronics.

Computer Graphics and Image Analysis: One of the most well-known applications of fractals is in computer graphics for generating natural-looking landscapes, clouds, and other virtual natural phenomena. Software that employs fractal generating algorithms can create highly detailed and realistic textures on limited computational budgets. In image analysis, fractals assist in compressing image data without significant loss of detail, capitalizing on the fact that many natural scenes exhibit fractal-like structures.

Environmental Science and Geography: Fractal patterns are also apparent in geographical formations such as coastlines and mountain ranges. This realization has led to the development of more accurate models for studying and predicting geographical and environmental phenomena. For instance, fractal geometry aids in modeling the spread of pollutants in diverse environmental systems, understanding the scaling laws of river networks, and predicting earthquakes and landslides.

Financial Markets: Interestingly, fractal mathematics has crossed into the realm of finance in analyzing stock market behaviors. Financial markets exhibit complex, dynamic behaviors that traditional models struggle to predict. Fractals help in building models that accommodate the market's inherent volatility and provide tools for analyzing risk more effectively.

Within each of these disciplines, fractal geometry bridges the gap between theoretical models and real-world phenomena by capturing the complexity of natural forms and processes. This ability not only enhances our understanding but also improves how we innovate and interact with our environment. Through these various applications, fractals continue to push the boundaries of what is achievable across scientific and technological landscapes, mirroring the detailed complexity that they so fundamentally represent.

3.7 Fractals in Art and Design

The presence of fractals in art and design highlights not only the aesthetic appeal of these structures but also underscores the interplay between science and art. Initially, artists were captivated by the intricate nature and seemingly endless complex patterns that fractals offer, even before the term 'fractal' was formally defined by Benoit Mandelbrot in 1975. This fascination has transcended traditional boundaries, influencing both mystic patterns in religious art and contemporary digital art forms.

Historically, certain elements of fractal geometry can be observed in African art, particularly in textiles, sculpture, and decoration. These patterns often exhibit self-similarity across different scales, a defining characteristic of fractals. Moving forward through different cultures and epochs, fractal elements can be identified in the artwork of various civilizations, inherently appealing to a universal aesthetic sensitivity.

In modern times, with the advent of digital technologies, artists have utilized fractal-generating software to create complex images that are based on mathematical algorithms. These tools have not only democratized the ability to produce patterns whose complexity was previously unattainable by hand but have also opened new frontiers in graphic design. Software like Apophysis or Ultra Fractal enables artists to explore intricate detail, create stunning visuals with dynamic color transformations, and manipulate scale with ease, offering a new palette and technique for artistic expression.

Examples abound, but one notable artist, Daniel Brown, uses fractal software to design unique works that are deeply influenced by patterns found in nature. His approach exemplifies how fractal algorithms can be tuned to mimic the natural growth processes of plants and other biological systems. Brown's work serves as a bridge connecting mathematical theory with tangible artistic creations that resonate with natural beauty.

Fractal art has not only been confined to two-dimensional canvases but has expanded into installations and architectural designs. Architects have used fractal principles to create structures with optimal light diffusion, natural airflow patterns, and aesthetically pleasing motifs that repeat at various scales. The result is buildings that are both functional and artistic, illustrating a practical application of fractals in large-scale designs.

To enhance engagement and understanding, consider an interactive exhibit that visually demonstrates how varying the parameters of a fractal algorithm changes the art produced. Viewers could adjust aspects such as iteration depth, transformation functions, and color schemes. This hands-on approach helps demystify the complexities behind fractal design and makes the mathematics behind it more accessible.

A key aspect of integrating fractals in design involves understanding their underlying mathematical structure. Fractals are characterized by self-similarity across scales, and this can be manipulated artistically to produce visually captivating patterns. To demonstrate, take the example of the Mandelbrot set, a famous fractal pattern. When zoomed into, the boundary of the Mandelbrot set reveals progressively intricate structures that echo the whole. Such characteristics can be instrumental in creating designs that have a balance between randomness and order, ideal for applications such as fabric prints, wallpaper designs, and even logos.

Indeed, the journey of fractals from abstract mathematical concepts to staple elements in art and design underscores a broader theme: that the boundaries between scientific inquiry and artistic endeavor are not merely blurred but mutually enriched. This fusion promises continued inspiration and innovation in both fields, as artists and scientists alike explore the depth and beauty of fractals.

3.8 Mathematical Challenges and Advances in Fractal Theory

Fractal theory presents a fascinating blend of visual aesthetics and mathematical depth, offering both challenges and advances that have reshaped our understanding of complexity within chaos. This evolution is punctuated by a series of significant breakthroughs in mathematics and computational science, as well as ongoing challenges that spur further innovation.

The computational generation of fractals, a critical element in exploring their properties, has advanced significantly with the adoption of increasingly powerful computers and algorithms. Initially, the iterative processes used to visualize fractals were computationally expensive, with systems often taking considerable time to produce detailed images. One major advance in this area has been the development of optimized algorithms such as the Fast Fourier Transform (FFT) applied in fractal image compression. This technique not only speeds up the generation process but also makes fractals more accessible and applicable in diverse fields such as digital imaging and telecommunications.

In mathematics, one of the seminal challenges has been the rigorous definition and measurement of fractal dimensions. The concept of dimension in fractals diverges from traditional geometric interpretations, as fractals exhibit complexity that varies with scale. Researchers have developed several methods to measure fractal dimension, such as the box-counting method, which counts the number of boxes of a given size needed to cover a fractal. These measurements, however, come with complications due to fractals' inherently intricate nature. The difficulty lies in determining theoretical bounds for fractal dimensions and validating these through practical computation, a challenge that is still not completely resolved.

Another significant mathematical challenge in fractal theory is related to probability and statistics—specifically, the modeling of random fractals. These structures form the basis of phenomena such as turbulent fluid flows and financial market models. The non-linear dynamics inherent in these systems make them difficult to predict and model. Advances in stochastic processes and chaotic dynamics have led to better predictive models, but accurately portraying the probability distributions within these realms remains a leading edge of current mathematical research.

From a theoretical standpoint, the intersection of fractal theory with other mathematical areas has been rich with development. The integration of fractal geometry with algebraic geometry, for instance, has uncovered new classes of objects whose fractal-like properties are intertwined with algebraic structures. This symbiosis between different domains of mathematics not only enriches each field but also opens up new avenues for thinking about and understanding complexity in nature.

3.8. MATHEMATICAL CHALLENGES AND ADVANCES IN FRACTAL THEORY

Diving deeper into visualization, the use of three-dimensional fractals has been an area of both challenge and advance. The creation and manipulation of 3D fractals require extensive computational resources and sophisticated software engineering. Recently, advances in computer graphics have allowed for real-time rendering of complex fractal landscapes, which are being used in virtual reality (VR) and augmented reality (AR) applications, providing users with an immersive experience of mathematical abstractions.

As we continue to explore fractals, their applications invariably lead back to their fundamental mathematical properties, and vice versa. This reciprocity between application and theory not only broadens the applicability of fractal concepts but also deepens our understanding of them. The journey through the complexity of fractals is symbolized not just by the answers we find along the way but more so by the quality and depth of the questions we learn to ask.

CHAPTER 3. FRACTALS: THE PATTERNS OF CHAOS

Chapter 4

Nonlinear Dynamics: Understanding the Rules

This chapter addresses the core principles of nonlinear dynamics, the branch of mathematics that deals with systems whose outputs are not directly proportional to their inputs. It explains key concepts, such as stability, bifurcations, and strange attractors, and provides insights into how these ideas elucidate the behavior of complex systems. Through analytical techniques and numerous examples, the chapter illustrates the fundamental rules that govern nonlinear systems and their unpredictable behaviors.

4.1 Introduction to Nonlinear Dynamics

Nonlinear dynamics is a fascinating realm of mathematics, profoundly instrumental in understanding the complex behaviors observed in natural and engineered systems. This field explores how systems react to inputs in ways that are not straightforwardly predictable or proportionally scalable. The implications of nonlinear dynamics stretch across multiple disciplines, including meteorology, engineering, economics, biology, and even philosophy, influencing how we understand causality and predictability in our surroundings.

One of the fundamental hallmarks of a nonlinear system is its sensitivity to initial conditions, often popularized as the "butterfly effect" in chaos theory. This concept suggests that a tiny disturbance in the initial state of a system can lead to vastly different outcomes, which illustrates the intrinsic unpredictability inherent in such systems. Mathematically, this effect is engendered by differential equations that do not possess solutions adhering to the superposition principle.

To delve a bit deeper, nonlinearity in differential equations implies that the governing equations of the system cannot be expressed merely as a linear combination of independent components. For example, if one considers the equation $\dot{x} = x^2 - t$ where $\dot{x}$ denotes the derivative of x with respect to time t, the non-linear term x^2 indicates that the rate of change of x not only depends on x but does so in a non-proportional manner. This non-linearity manifests as complex dynamics when the system evolves over time.

Another intriguing aspect of nonlinear dynamics is the presence of bifurcations, which are points in the parameter space of the system at which the qualitative nature of the dynamical system changes. For instance, a simple increase in the temperature of a chemical reactor can shift the system from a steady state into oscillatory behavior.

Nonlinear dynamics embrace a variety of behaviors including periodic orbits, quasiperiodic routes to chaos, and chaotic dynamics, each constituting an integral part of the study. The Lorenz attractor, for example, representing a simplified model of atmospheric convection, exhibits a chaotic flow characterized by continuous dependency on initial conditions and an aperiodic long-term behavior.

Tools such as phase portrait analysis, Lyapunov exponents, and Poincaré maps are indispensable in studying these systems. These tools allow scientists and mathematicians to visualize complex behavior, estimate system stability, and map out the trajectories in the phase space, respectively, providing crucial insights into the underlying dynamics.

The dance of numbers that is nonlinear dynamics offers more than just an understanding of mathematical equations; it unfolds stories about the interconnectedness and unpredictability of our world. As we peel layer by layer, the structure and function of complex systems become less mystifying. The journey through this elaborate mathematical landscape is not merely about solving equations, but rather about piecing together the mosaic of real-world phenomena that these equations describe.

4.2 Contrasting Linear and Nonlinear Systems

In the exploration of dynamic systems, understanding the distinction between linear and nonlinear systems provides crucial insight into why some systems are predictable and others manifest chaotic behavior. Linear systems, where the output is directly proportional to the input, adhere to the principle of superposition. This principle states that the sum of the responses caused by individual inputs is equal to the response caused by the sum of those inputs. Mathematically, if a system is described by $f(x)$, and it is linear, then for any two inputs a and b and any scalars c and d, the function satisfies:

$$f(ca + db) = cf(a) + df(b).$$

The predictability and simplicity of linear systems arise from this property, leading to solutions that are straightforward to calculate using traditional algebraic methods.

Nonlinear systems, in contrast, do not satisfy the superposition principle. These systems are described by equations in which the change rules involve powers, products, or transcendental functions of the variables that do not merely scale with increases in input. For instance, a nonlinear differential equation may look something like:

$$\frac{d^2 x}{dt^2} + ax^3 + b = 0,$$

where x represents the system state and a and b are parameters. Here, the term ax^3 is a clear indicator of nonlinearity due to its cubic relationship with x.

Considering how these different systems respond to similar initial conditions can be quite revealing. In linear systems, similar initial conditions yield similar trajectories. For nonlinear systems, however, even slight variations in initial conditions can lead to vastly different outcomes—a characteristic known as sensitive dependence on initial conditions, which is a hallmark of chaotic systems.

Furthermore, the techniques employed in studying these systems differ significantly. For linear systems, solutions

4.2. CONTRASTING LINEAR AND NONLINEAR SYSTEMS

often involve straightforward operations such as matrix inversions or solving linear differential equations. Techniques such as eigenvalue analyses are also common, allowing for a decomposition of the system response into modes that can individually be analyzed.

On the other hand, studying nonlinear systems often requires iterative methods, numerical simulations, and perturbative analysis, which approximate the behavior of the system under small deviations from typical values. For example, consider the nonlinear spring equation:

$$\frac{d^2x}{dt^2} = -kx - \alpha x^3.$$

To solve this equation, one might use a method like perturbation theory to deal with the cubic term, treating it as a small correction to the linear term.

Another insightful aspect is the visualization of system behavior via phase space diagrams. For linear systems, these diagrams typically consist of straight lines or ellipses. In contrast, nonlinear systems may display complex and intricate patterns indicative of multiple equilibria or strange attractors—phenomena that would be impossible in linear systems.

As practitioners and theorists delve deeper into nonlinear dynamics, they find that many real-world systems are inherently nonlinear. Weather systems, economic markets, biological phenomena, and many mechanical systems display behaviors that can only be accurately modeled with nonlinear mathematics. This realization has profound implications for our ability to predict, control, and optimize these systems.

Thus, a comprehension of why nonlinear systems deviate so markedly from their linear counterparts paves the way for a richer understanding of complex patterns we observe in both nature and technological applications. This paradigm shift encourages innovative thinking and advanced methodologies in tackling problems that defy linear solutions, fostering advancements across science and engineering disciplines.

4.3 Key Principles of Nonlinearity

Nonlinear dynamics, fundamentally fascinating yet intricate, revolves around the concept that simple rules can generate unpredictable outcomes. It is paramount in our exploration to unravel some key principles that highlight the characteristic features of nonlinearity.

Sensitivity to Initial Conditions: One of the most remarkable aspects of nonlinear systems is their extreme sensitivity to initial conditions, often popularized as the "butterfly effect." This principle posits that tiny variations in the starting point of a system can lead to vastly different outcomes, making long-term prediction nearly impossible in such systems. Mathematically, this is depicted by the divergence of trajectories in phase space, which can be quantified using Lyapunov exponents. A positive Lyapunov exponent typically indicates chaotic behavior and, thus, high sensitivity to initial conditions.

$$\lambda = \lim_{t \to \infty} \frac{1}{t} \log \frac{|\delta \mathbf{x}(t)|}{|\delta \mathbf{x}(0)|}$$

where $\delta \mathbf{x}(0)$ is a small perturbation in the initial state, and $\delta \mathbf{x}(t)$ is its evolution over time.

Feedback Loops: Feedback loops are pivotal in nonlinear systems. They can amplify or moderate the effects of a system's inputs through positive or negative feedback. Positive feedback reinforces the input signal, potentially leading to exponential growth or runaway effects. On the other hand, negative feedback tends to resist changes, promoting system stability. These feedback mechanisms are pivotal in environmental, economic, and biological systems, dictating patterns such as population growth or economic cycles.

Phase Space and Attractors: The concept of phase space provides a powerful framework for visualizing the entire state of a system. Within this space, trajectories represent possible states through time. Nonlinear systems often possess complex attractors known as "strange attractors" when they demonstrate chaotic behavior. These attractors have a fractal dimension and are sensitive to initial conditions. Essentially, they pull trajectories into a set pattern, though the paths taken by different initial conditions may appear random.

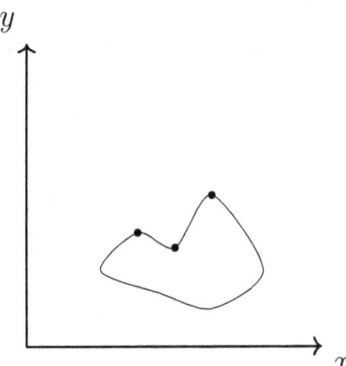

Figure 4.1: Illustrative example of a strange attractor in a two-dimensional phase space.

Scaling and Self-similarity: Nonlinear systems often exhibit scaling properties and self-similarity over different scales. This fractal nature implies that one can observe similar patterns at various scales of measurement. This scaling property is not only a philosophical curiosity but provides practical tools for analyzing data across different scales, particularly useful in fields like meteorology and financial markets.

To fully appreciate these principles in action, one might consider the logistic map, a deceptively simple nonlinear equation that exhibits rich chaotic behavior as its parameters are varied. The logistic map adequately demonstrates sensitivity to initial conditions, feedback loops effect through population regulation under limited resources, and dramatic shifts in dynamics based on slight changes in the growth rate parameter.

Understanding these principles enables analysts and researchers to grasp the profound complexity hidden within seemingly straightforward systems. This understanding further entrails the potential to manipulate or predict the behavior of complex systems ranging from weather systems to economic markets, making the study of nonlinearity not just academically intriguing but also practically indispensable.

4.4 Tools for Analyzing Nonlinear Dynamics

The exploration of nonlinear dynamics depends heavily upon a variety of mathematical and computational tools that enable us to analyze and predict the behavior of complex systems. Primarily, these tools can be classified into analytical methods, numerical simulations, and experimental tools. Each category plays a pivotal role in helping us unveil the layers of complexity inherent in nonlinear systems.

Phase Space Analysis: One of the most enlightening tools in our arsenal is phase space analysis. Phase space provides a multi-dimensional space in which all possible states of a system are represented, with each axis corresponding to one of the system's variables. Thus, for a dynamical system with two variables, the phase space is a plane. By plotting the trajectory of the system's state over time, we can visualize how the system evolves. For instance, the presence of limit cycles or strange attractors can be discerned from this representation. Using TikZ in LaTeX, we could represent a simple phase diagram of a hypothetical system with a limit cycle as follows:

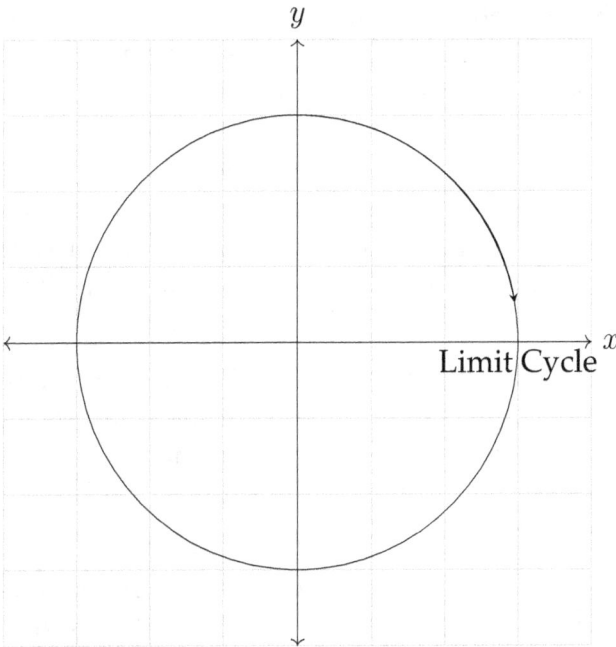

Lyapunov Exponents: Another critical tool is the calculation of Lyapunov exponents, which quantify the rate at which nearby trajectories in the phase space diverge from each other. A positive Lyapunov exponent indicates chaotic behavior, as small differences in initial conditions grow exponentially over time, leading to drastically different outcomes. This sensitivity to initial conditions is a hallmark of chaotic systems.

Bifurcation Analysis: As systems undergo changes in parameters, their qualitative behavior can change significantly. Bifurcation analysis helps in understanding these transitions by mapping out the bifurcations or qualitative changes in the system's dynamics as parameters are varied. This tool can reveal the thresholds at which a system's behavior shifts from stable to chaotic, or from oscillating to steady state.

Numerical Simulations: With advancements in computing, numerical simulations have become indispensable. The most famous among these is the Runge-Kutta method, particularly useful for solving differential equations numerically when analytical solutions are intractable. Numerical simulations allow for approximations of a system's future behavior based on its current and historical states, providing valuable insights into dynamics that are complex to solve analytically.

Poincaré Sections: This method involves taking snapshots of a higher-dimensional system at regular intervals (or slices) to reduce it to a lower-dimensional system. This analysis simplifies understanding complex orbits by dissecting them into easier-to-analyze intersections.

Each tool not only provides a different lens to understand the intricate behaviors of nonlinear systems but also complements the other tools to provide a comprehensive understanding of such dynamics. Leveraging these tools effectively requires a blend of mathematical insight and practical intuition, where theoretical knowledge meets real-world application.

4.5 Stability and Instability in Systems

In nonlinear dynamics, the concepts of stability and instability are paramount in understanding how systems behave under various conditions. Stability in a system implies that the system will return to its equilibrium state after a small disturbance. Conversely, instability suggests that the system will diverge from its equilibrium state under perturbation. These concepts are vital in predicting the behavior of many natural and engineered systems.

Stability can be characterized as either local or global. Local stability refers to system behavior near an equilibrium point, where small perturbations do not lead to significant changes or lead the system to return to its equilibrium state. Global stability, however, considers the whole space in which the system operates, ensuring that all possible behaviors from any initial state will eventually converge to an equilibrium.

To analyze stability, one often examines the eigenvalues of the Jacobian matrix at a given fixed point. For a system described by the differential equation $\dot{x} = f(x)$, where x is a state vector and f represents the system dynamics, the Jacobian matrix J, defined as $J = \frac{\partial f}{\partial x}$, is computed at the equilibrium point. If all the eigenvalues of J have negative real parts, the system is locally stable at that point. However, if any eigenvalue has a positive real part, the system is locally unstable.

Instability in nonlinear systems is not merely the opposite of stability; it can exhibit complex behaviors. For instance, a stable system affected by a small disturbance will settle back into a steady state, but an unstable system might enter into a new type of behavior entirely, such as chaos or oscillation among multiple states. These behaviors are crucial for understanding phenomena such as weather patterns, where a seemingly stable system can suddenly change into a turbulent state.

Consider the nonlinear system given by:

$$\dot{x} = \sigma(y - x), \quad \dot{y} = rx - y - xz, \quad \dot{z} = xy - bz$$

known as the Lorenz system. Here σ, r, and b are parameters that dictate the system's behavior. Analysis of this system shows that for certain values of these parameters, the system exhibits chaotic behavior, a form of instability that is highly sensitive to initial conditions.

The use of phase portraits and bifurcation diagrams provides further insight into system stability and instability. A phase portrait is a graphical presentation that shows all possible trajectories of the system in the phase space, giving us a visual representation of the stability characteristics of different states. Bifurcation diagrams, on the other hand, show how the qualitative behavior of equilibrium points changes as a function of system parameters.

For instance, consider producing a basic bifurcation diagram for a one-dimensional system:

$$\dot{x} = r + x - x^3$$

Using software like MATLAB or Python's Matplotlib library, one can plot how the equilibrium points vary with r. This diagram would illustrate shifting stabilities and emerging bifurcations, which mark transitions from stability to instability as r changes.

To conclude this discussion, it is important to appreciate how deeply stability and instability are intertwined with the fundamental behaviors observed in nonlinear systems. Their analysis not only provides insights into predictable patterns but also helps in understanding and preparing for the unpredictable, often chaotic, behaviors that can arise in complex systems.

4.6 Bifurcations: Changes in System Behavior

Bifurcations are critical junctions in the behavior of nonlinear systems where a small change in a parameter, such as the system's input or environmental conditions, causes a qualitative or topological change in its dynamics. This concept is pivotal in understanding how complex systems can undergo sudden transitions from stable to chaotic behavior—or between different forms of stability. A deeper dive into the types of bifurcations provides insight into predicting and managing such transitions in practical scenarios.

Types of Bifurcations The study of bifurcations classifies them typically into several main types, each characterized by the nature of the parameter change and the resulting system behavior. The most common types include saddle-node, transcritical, pitchfork, and Hopf bifurcations.

- **Saddle-node Bifurcation:** This occurs when two

equilibrium points of the system—typically one stable and one unstable—collide and annihilate each other as a parameter is varied. This can lead to the system suddenly losing stability, an event commonly observed in electrical circuits and ecological models.

- **Transcritical Bifurcation:** In this type, an equilibrium point of the system changes its stability as it passes through another equilibrium. This sort of behavior is noted in biological systems where increasing population density might shift an equilibrium from stable to unstable, leading to sudden population declines or explosions.

- **Pitchfork Bifurcation:** Distinguished by the symmetry in its ordinary differential equations, pitchfork bifurcation can be supercritical (giving rise to stable solutions) or subcritical (leading to unstable solutions). The branching nature of this bifurcation portrays scenarios such as magnetization in physics where the magnetic state can suddenly change direction.

- **Hopf Bifurcation:** Involves a pair of complex conjugate eigenvalues crossing the imaginary axis to create or annihilate a limit cycle. This type is often associated with systems displaying periodic behavior which can suddenly emerge or disappear with changes in parameters.

Analytical Tools Analyzing bifurcations requires a range of mathematical tools, with normal forms and center manifold theory being particularly powerful in simplifying complex dynamical systems to their most

4.6. BIFURCATIONS: CHANGES IN SYSTEM BEHAVIOR

essential components. These reduced models retain the critical features necessary to study bifurcations without the complications introduced by higher dimensions and non-essential dynamics.

$$\dot{x} = rx - x^3$$

Above equation presents an example of a normal form for a pitchfork bifurcation where r signifies a bifurcation parameter. Variations in r highlight transitions between no equilibria and the existence of three equilibria, of which two are stable and one unstable when $r > 0$.

Implications in Applications Understanding the types of bifurcations and their underlying mechanisms allows scientists and engineers to predict sudden changes in system behaviors—and potentially control them. In climate science, anticipating tipping points in climate models can inform better decision-making for environmental policies. In mechanical engineering, controlling parameters to avoid undesirable bifurcations ensures safer operation of machinery.

As diverse as bifurcations are, their study embodies the crux of exploring nonlinear systems, revealing the myriad ways by which stability can be lost and new behaviors can emerge. Navigating through this complex landscape not only highlights the intricacies of dynamic systems but also reflects a broader theme: the delicate balance of stability and change that pervades the universe. The ability to predict and manage these changes is not just about understanding or controlling chaos, but about harnessing it—a challenge that continues to inspire researchers and practitioners across disciplines.

4.7 Chaos in Nonlinear Systems

Chaos is a fundamental aspect when discussing nonlinear dynamics, a term that often evokes images of unpredictability and random behaviors that are difficult, if not impossible, to predict precisely over long time periods. This misconception, that chaos is merely disorder or randomness, misses the nuance that chaotic systems are deterministic; their future behavior is fully determined by their initial conditions, albeit sensitive to them.

To delve deeper into the concept, consider the Lorentz system, a classic example of a chaotic system. The Lorentz attractor arises from a set of differential equations modeling convection rolls in the atmosphere:

$$\frac{dx}{dt} = \sigma(y - x),$$
$$\frac{dy}{dt} = x(\rho - z) - y,$$
$$\frac{dz}{dt} = xy - \beta z.$$

with σ, ρ, and β being parameters representing physical properties of the system. For certain values of these parameters, the system exhibits chaotic behavior.

The distinctive feature of chaos is the sensitive dependence on initial conditions, commonly known as the butterfly effect. To illustrate, let's plot how two trajectories from closely spaced initial conditions diverge over time using TikZ in LaTeX:

This figure clearly displays how minute differences in starting conditions can lead to drastically different outcomes, a characteristic behavior of chaotic systems.

Another significant aspect of chaotic systems within

4.7. CHAOS IN NONLINEAR SYSTEMS

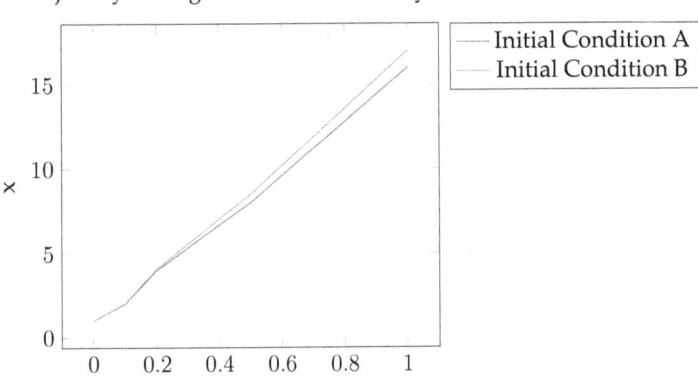

Figure 4.2: Initial condition sensitivity in the Lorentz System.

nonlinear dynamics is the presence of strange attractors. These are attractors that have a fractal dimension and do not settle down to a single state, but rather evolve over time into what appears to be a random state. This complex set where the dynamics transpire indicates continuing motion without repetition. For instance, the Lorentz attractor previously described is one such strange attractor. It does not repeat and it plots a unique pattern in coordinate space.

Further analyzing chaotic behaviors, techniques like Lyapunov exponents are used which quantitatively measure the average rate of separation of infinitesimally close trajectories. A positive Lyapunov exponent typically provides a quantitative indication of chaos, suggesting that small changes in initial conditions will lead to exponential divergence of trajectories.

The unraveling of chaotic behaviors in nonlinear systems aids significantly in understanding natural phenomena ranging from weather patterns to the rhythms of the heart. By studying these systems, scientists can begin to predict aspects of their behavior over short timescales, which is crucial for planning and simulations in various fields.

Acknowledging the implications of chaos theory not only transforms our theoretical approaches but also enhances our appreciation of the natural complexity present in the world around us.

4.8 Case Studies: Nonlinear Dynamics in Action

Let us explore how the theoretical frameworks of nonlinear dynamics apply to real-world scenarios through multiple case studies. These examples span across various disciplines, emphasizing the ubiquity and importance of nonlinear dynamics in understanding complex systems.

Weather Prediction and Atmospheric Dynamics: One of the most compelling examples of nonlinear dynamics at work is in weather prediction. The atmosphere is a chaotic system characterized by complex interactions among various factors such as air temperature, pressure, and moisture content. Edward Lorenz, one of the pioneers in chaos theory, discovered what is now known as the Lorenz attractor, a set of chaotic solutions to a simplified model of atmospheric convection. This fundamentally changed our approach to meteorology, illustrating that tiny changes in initial conditions can lead to drastically different outcomes, making long-term weather forecasting an immense challenge.

4.8. CASE STUDIES: NONLINEAR DYNAMICS IN ACTION

Abstract Lorenz Attractor

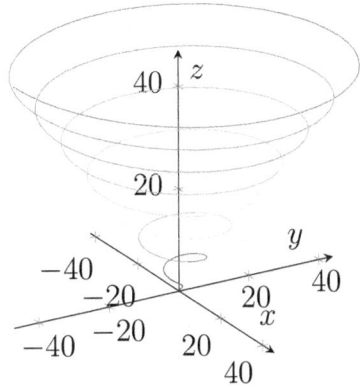

Population Dynamics: Another realm where nonlinear dynamics plays a critical role is in ecology, particularly in understanding population dynamics. The logistic map, one important model in this area, describes how populations evolve over time, factoring in the rates of reproduction and starvation. This model can lead to various types of behavior, from stable fixed points to chaotic fluctuations depending on the reproductive rate. This has significant implications for ecological management and conservation efforts, helping biologists develop strategies to maintain species diversity and prevent extinction.

Economic Models: Nonlinear dynamics also finds application in economics, especially in the modeling of market cycles and financial crises. Nonlinear models can capture the complex behaviors seen in economic systems, where traditional linear models fail. For instance, the interplay between inflation rates, consumer behavior, and regulatory changes can lead to unexpected economic fluctuations and crises, which can be better understood through nonlinear economic models.

Example of a Nonlinear Economic Model

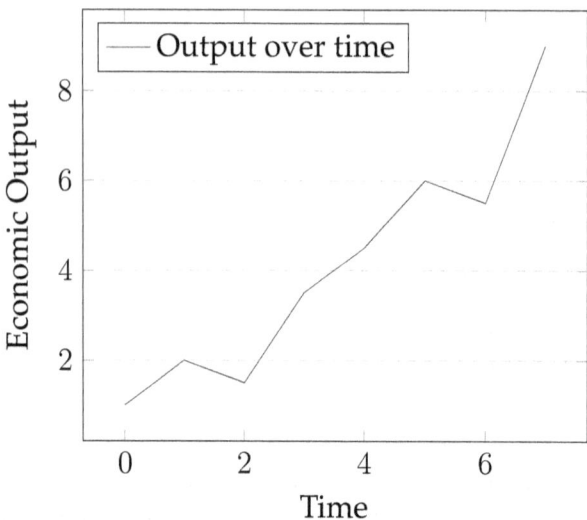

These cases highlight not only the breadth of applications for nonlinear dynamics but also underscore the need for robust nonlinear models that can enhance our understanding and predictive capabilities in varied fields. By studying these systems through the lens of nonlinearity, we gain insights that are fundamentally impossible to achieve through linear analysis alone.

As we delve into these diverse studies, it becomes evident that the world around us behaves not just in increments and linearity, but often exhibits sudden changes, unpredictability, and complexity that challenge our conventional understandings. Exploring these phenomena through nonlinear dynamics not only broadens our scientific horizons but also provides practical tools to navigate and influence these systems effectively.

In this section, detailed implementations of real-world scenarios underscore the critical importance and widespread applicability of nonlinear dynamics across various fields.

4.8. CASE STUDIES: NONLINEAR DYNAMICS IN ACTION

Through structured observations and mathematical modeling, we are continually discovering how deep the roots of nonlinearity extend into the fabric of natural and human-made systems.

CHAPTER 4. NONLINEAR DYNAMICS: UNDERSTANDING THE RULES

Chapter 5

Chaos in Nature: Weather, Ecology, and Beyond

This chapter investigates the presence and implications of chaos theory within natural phenomena, specifically focusing on weather and ecological systems. It discusses how chaotic behavior is not only prevalent but plays a predictive role in understanding natural processes from precipitation patterns to population dynamics. By exploring case studies and scientific findings, the chapter highlights the instrumental role of chaos theory in advancing the study of environmental science and managing natural resources more effectively.

5.1 Chaos in Weather Patterns

Delving into the subtleties of chaotic dynamics in weather patterns unveils a complex picture where the seemingly random occurrences of weather events stem from nonlinear interactions within the Earth's atmosphere. Unlike

CHAPTER 5. CHAOS IN NATURE: WEATHER, ECOLOGY, AND BEYOND

other systems discussed in this book, the atmosphere's enormity and variability make it a quintessential example of chaos on a grand scale.

The atmospheric system is primarily driven by the interplay between Earth's rotation, the sun's heat, and the properties of air and water. This system is sensitive to initial conditions, a fundamental aspect of chaos theory. The famous 'butterfly effect,' which suggests that the small flap of a butterfly's wings can ultimately lead to a tornado weeks later on another continent, is emblematic of this sensitivity.

Mathematically, weather forecasting relies heavily on numerical models that solve complex differential equations. These equations represent fluids dynamics and thermodynamics under various conditions, incorporating variables such as temperature, pressure, humidity, and wind velocity. A characteristic equation relevant in this domain is the Lorenz system, formulated by Edward Lorenz in 1963, which simplified the atmosphere's dynamics into a system of three nonlinear differential equations. The solutions to these equations exhibit what is known as Lorenz attractors, providing a visual representation of chaos.

As practical as these models are, their predictive accuracy diminishes sharply as the forecast period extends.

The initial condition's slightest error grows exponentially due to the chaotic nature of the system. To illustrate, if a weather model has a minor error in its initial humidity levels, this small discrepancy can lead to significantly different weather outcomes over time. This exponential growth of errors sets practical limits on reliable forecasts.

Supercomputers play a crucial role in processing the enormous amount of data required for weather prediction and in performing simulations based on atmospheric models. They iterate thousands of potential scenarios in ensemble forecasting, a technique that attempts to address uncertainties and chaos by running many forecasts with slightly varied initial conditions.

Despite advances in technology and methodology, predicting weather patterns remains a challenge due to the inherent chaotic nature of the atmosphere.

As technology progresses and more refined data are procured, improved prediction accuracy will likely be achieved by embracing rather than resisting the chaotic nature of weather systems. This evolving understanding enables us to better anticipate and mitigate the impacts of increasingly erratic climatic phenomena.

5.2 Ecological Systems and Chaos

The exploration into ecological systems reveals a fascinating tapestry where chaos theory not only applies but substantially enhances our understanding of ecosystem dynamics. A prime example of chaos within ecological systems can be seen in predator-prey relationships, where traditional models like the Lotka-Volterra equations introduce the fundamentals of oscillatory dynamics that may evolve into chaotic patterns under certain conditions.

To delve deeper, consider a basic predator-prey model modified to include more realistic features such as limited resource availability and disease factors among prey. The mathematical representation of such a system, while initially straightforward, quickly becomes complex as additional factors are included. Charting prey population (P_t) against predator population (N_t) over time might typically present a periodically stable cycle. However, varying the growth rate, carrying capacity, or other environmental influences can lead to erratic and unpredictable behaviors typical of chaotic systems.

Visualization through Phase Space Diagrams

Employing phase space diagrams provides a vivid illustration of how these dynamical systems can shift from predictable to chaotic behavior. In such diagrams, each point represents the state of the system at a given time, with axes typically showing population sizes of prey and predators. A stable ecological system might display a closed loop or orbit, indicating regular cycles in population sizes. Contrastingly, in a chaotic regime, these orbits decompose, covering a broader area of the phase space erratically, which signifies the sensitive dependence on initial conditions—a hallmark of chaos.

5.2. ECOLOGICAL SYSTEMS AND CHAOS

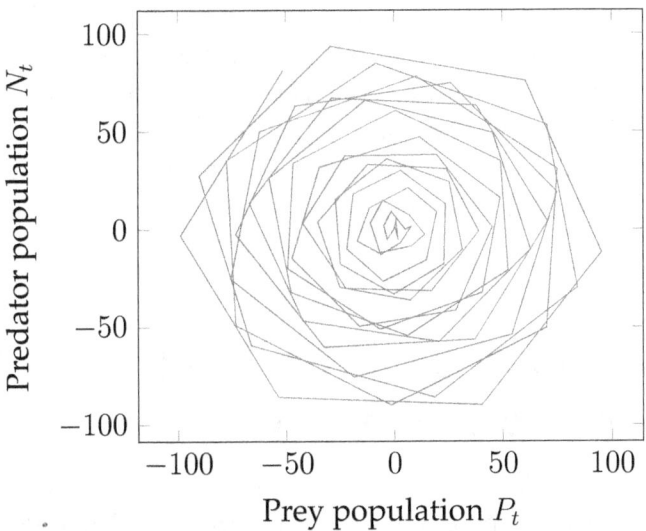

Phase Space Diagram of a Predator-Prey System

Implications for Conservation and Resource Management

The chaotic attributes found in these systems also bring significant implications for conservation efforts and resource management. Policies based on average conditions or static models may fail because they do not account for the underlying variability and potential for sudden changes in the system caused by chaos. Adaptive management strategies that recognize and plan for potential chaotic fluctuations can lead to more robust approaches to biodiversity conservation and sustainable resource use.

Additionally, converting our theoretical understanding of chaos in ecological models into actionable insights entails employing numerical simulations and continuations of observable ecological data. These methodologies allow researchers to anticipate potential chaotic shifts and bifurcations in ecosystem behaviors that would affect management practices.

This integration of chaos theory into ecological studies enriches our comprehension of the complex interactions and unexpected behavior patterns among various species within an ecosystem, thus preparing us to better manage and conserve our natural environment amidst increasing global changes and challenges.

5.3 Geophysical Phenomena and Nonlinear Dynamics

One of the most emphatic demonstrations of chaos theory can be observed in geophysical phenomena, where nonlinear dynamics govern systems ranging from mantle convection in geology to fluid dynamics in oceanography. Nonlinear systems are those in which the output is not proportional to the input, often resulting in complex, unpredictable behavior. This nonlinearity is intrinsically linked to the chaotic patterns observed in Earth's geophysical processes.

Starting with the atmospheric sciences, we consider the example of turbulent fluid flows, which are quintessentially chaotic. Turbulence in the atmosphere and oceans is driven by nonlinear interactions within the fluid motions. These interactions often lead to what is known in mathematics as sensitivity to initial conditions, or more commonly known as the butterfly effect. This principle posits that small changes in the initial state of a system can lead to vastly different outcomes, a concept that is crucial for understanding weather systems and climate models.

5.3. GEOPHYSICAL PHENOMENA AND NONLINEAR DYNAMICS

In oceanography, chaos manifests in the form of ocean currents and eddies. The dynamic systems that describe ocean circulation are characterized by their turbulent, unpredictable behavior. Mathematical models such as the Navier-Stokes equations, which are used to predict fluid movement, are inherently nonlinear. When attempting to solve these equations under realistic oceanic conditions, minute variations in input can significantly alter the outcome, thus demonstrating chaotic behavior. One can model these scenarios using the `pdfplots` package in LaTeX to visualize how small changes in initial conditions can lead to different trajectories in a system described by Navier-Stokes equations:

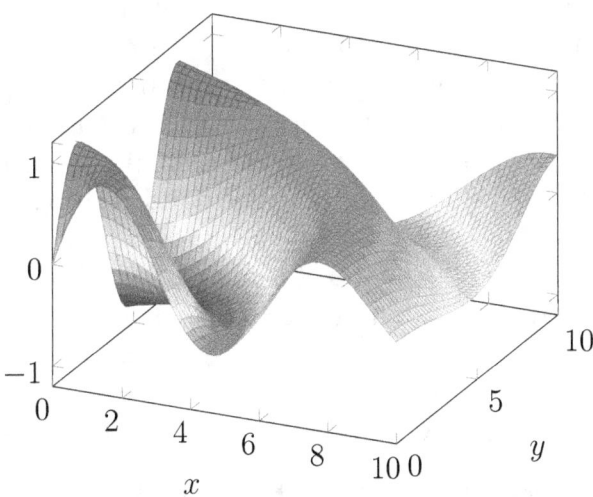

Furthermore, nonlinear dynamics plays a significant role in understanding geological changes and processes such as earthquakes and volcanic eruptions. For instance, the theory of plate tectonics, central to explaining geological phenomena, involves complex interactions amongst earth's plates that can exhibit chaotic behavior. Earthquake aftershock sequences and volcanic eruption forecasting are examples where chaos theory has been applied to predict seemingly random events with greater accuracy.

In addition to terrestrial applications, chaos and nonlinear dynamics also find their place beyond Earth, in broader astrophysical contexts. Astrophysical systems often involve multiple interacting bodies subject to gravitational forces leading to chaotic motion. The three-body problem in celestial mechanics is a classic example where predictive modeling becomes challenging due to the system's sensitivity to initial conditions.

The integration of chaos theory into geophysical research has provided significant insights that contribute to our understanding of natural systems and their behaviors. This approach has not only improved the accuracy of predictions in weather and geological forecasts but also deepened our understanding of the universe's fundamental laws as applied to complex dynamical systems. As our computational capabilities continue to expand, so too will our ability to model these systems with greater precision, potentially revealing new layers of complexity within the chaotic webs of natural phenomena.

5.4 Biological Chaos: From Heartbeats to Neural Networks

Delving into the realm of biological systems, it becomes readily apparent that chaos is not only an inherent characteristic but also a crucial component of their functionality. The inherent unpredictability in the timing of heartbeats and the firing of neural networks exemplifies chaotic dynamics at play. This section will explore how chaos theory has fundamentally reshaped our understanding of these vital biological processes, providing insights into their complex behavior which traditional linear models failed to explain adequately.

Starting with human heartbeats, a phenomenon known to many through experiences of palpitations or racing hearts, it is easy to assume such behaviors are merely symptomatic. However, under the lens of chaos theory, these fluctuations are more than irregularities; they are indicative of an underlying chaotic system. The heart's pacemaker cells generate electrical signals in a seemingly disordered manner which actually displays fractal properties—a hallmark of chaos. To understand this, researchers have analyzed inter-beat intervals using nonlinear dynamic tools such as phase space plotting and Lyapunov exponents. These studies reveal that the heart operates in a chaotic regime, which surprisingly aids in its efficiency and robustness. Healthy hearts demonstrate a greater degree of chaos compared to those of patients with certain types of heart diseases, who have less variability and more predictable patterns in their heartbeats.

Switching our focus to neural networks, these are even more complicated due to their extensive interconnectivity and ability to adapt and learn. Neurons exhibit firing patterns that, at first glance, seem stochastic and unpredictable. However, these patterns are the product of intricate neural circuits that dynamically change and adapt. The concept of 'neural chaos' is an active area of research which posits that the chaotic nature of neural activity allows for flexible responses to external stimuli, facilitating learning and memory formation. This chaos is characterized by a high sensitivity to initial conditions, which in the context of neural networks, means that even minor changes in synaptic input can lead to vastly different output patterns. This property is crucial for the brain's ability

CHAPTER 5. CHAOS IN NATURE: WEATHER, ECOLOGY, AND BEYOND

to encode and process the vast array of sensory inputs it constantly receives.

To illustrate these concepts, consider the application of the *tikz* package to plot a simple phase space diagram of a chaotic neural model. The diagram might look something like this:

Phase Space Plot of Neural Firing Patterns

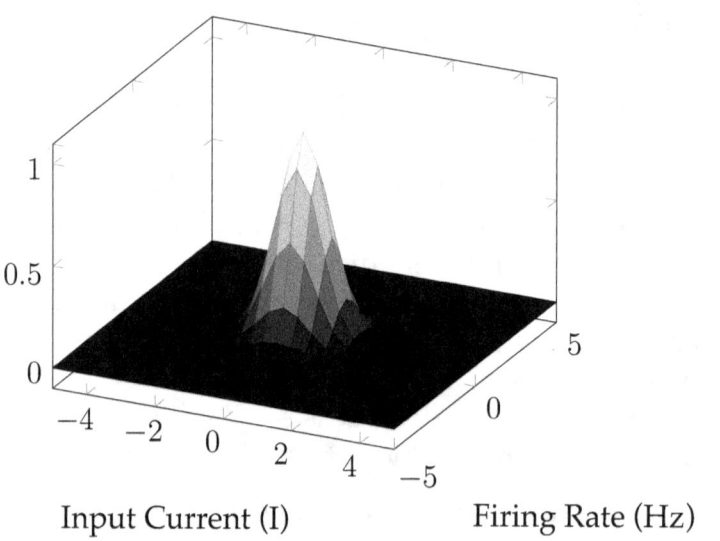

Input Current (I)   Firing Rate (Hz)

This hypothetical plot does not represent specific real-world data but is instead a symbolic representation. In true research, such diagrams would be derived from empirical data, showcasing clusters, loops, or spirals characteristic of chaotic attractors, providing insight into the operational modes of neural activity.

Understanding biological chaos has significant implications for medical science, particularly in the development of diagnostic and therapeutic tools. For heart diseases, it could mean designing treatments that aim to restore the healthy chaotic patterns rather than merely normalizing the rhythm. For neurological conditions, leveraging chaos might aid in devising interventions that help reestablish optimal chaotic states, thereby enhancing brain functionality.

With these explorations into the chaotic nature of heartbeats and neural networks, we recognize a profound shift from viewing these patterns as mere anomalies to understanding them as integral to physiological effectiveness and health resilience. It opens up not just a set of challenges but also a vista of opportunities for medical science—highlighting yet again how chaos theory continues to illuminate pathways toward comprehending and harnessing complexity in life sciences.

5.5 Impact of Chaos Theory on Environmental Science

Chaos theory, a principle that describes the inherent unpredictability of systems governed by deterministic laws that exhibit sensitive dependencies on initial conditions, provides a foundational framework for approaching environmental science. Contrary to expectations that might come with the term "chaos," this theory does not imply randomness, but rather unpredictability amidst underlying order, significantly affecting how environmental systems are studied and managed.

Understanding the dynamics of chaos theory in environmental science begins with the appreciation of how minute variations in initial conditions can lead to drastically different outcomes. This sensitivity, often referred to as the "butterfly effect," manifests itself clearly in climate models and ecosystem simulations where small changes in input parameters can result in widely varied environmental responses.

For instance, let us delve into climate modeling. Traditional models used linear approximations to predict weather patterns; however, these models could not accurately capture the complexity and dynamism inherent in atmospheric conditions. By incorporating algorithms that align with chaotic behavior, scientists now develop models that can better encapsulate the fluctuating nature of climate systems. These enhanced models are critical for improving weather forecasts and preparing more effectively for climate-related disasters.

Model Type	Predictive Time Frame	Accuracy
Traditional Linear Models	Short-term (1-3 days)	Moderately High
Chaotic Nonlinear Models	Mid-term (1-2 weeks)	High under certain conditions
Long-term Climate Predictions	Several years	Varies widely

Ecological systems also exemplify chaotic behavior. The population dynamics of species, for example, can exhibit chaotic patterns where small differences in birth rates, mortality rates, or seasonal changes can lead to surprisingly different outcomes in population sizes over time. By employing mathematical models accentuating these chaotic principles, ecologists have improved their understanding of species interactions, migration patterns, and the impact of human disturbances on wildlife and habitats.

In environmental resource management, chaos theory has revolutionized conservation strategies. Adaptive management is one such strategy that recognizes the chaotic nature of ecosystems and therefore implements a dynamic, iterative decision-making process in managing natural resources. Instead of setting rigid, long-term plans based on static models, resource managers now use short-term, flexible strategies that can be modified as more is learned about the system's responses to management actions.

The implications of chaos theory reach further into risk assessment and disaster preparedness plans. By accounting for chaotic variability, governments and organizations can better design their strategies to handle unexpected changes in environmental conditions. This is crucial for developing more robust societal responses to environmental hazards and disasters.

Chaos theory offers a paradigm shift in understanding and managing the natural environment. It accentuates the importance of detailed observation and adaptive strategies that respond to the inherently unpredictable behavior of complex systems. By embracing the insights provided by chaos theory, environmental science can advance in a way that is more aligned with the natural world's dynamism.

5.6 Predicting and Managing Natural Disasters

Within the domain of natural disasters, chaos theory's contributions are paramount, especially when it concerns prediction and management. Traditional approaches to natural disaster prediction and management often rely on linear models, which assume a direct, proportionate cause and effect. However, numerous systemic variables in natural phenomena are interlinked in a non-linear fashion, making these traditional methods insufficient. Chaos theory allows for a more nuanced understanding, focusing on sensitivities to initial conditions and the inherent unpredictability of systems at certain scales.

To exemplify, consider the case of earthquake prediction. Earthquakes are the product of complex interactions within the earth's crust. Despite extensive research, their precise prediction continues to elude scientists. Here, chaos theory aids in understanding the underlying dynamics that lead to seismic activity. Researchers are exploring how minor variations in tectonic stress and geological faults might cascade into significant earthquakes. By modeling these dynamics using non-linear mathematics, scientists can identify potential precursors to larger seismic events.

Similarly, in hurricane forecasting, chaos theory has revolutionized predictions by focusing on the iterative effects of small-scale meteorological elements – such as temperature variations in the ocean surface – which can drastically alter a storm's path and intensity. Advances in computational models, which incorporate principles of chaos and non-linear dynamics, are significantly improving the accuracy of hurricane tracks and intensity forecasts. For instance, the use of ensemble forecasting, where a multitude of potential scenarios is simulated, acknowledges the chaotic nature of weather systems by presenting a range of possible outcomes, rather than a single definitive prediction.

Fire dynamics in wildland fire management also demonstrate chaotic behavior, particularly in how small changes in wind speed and direction can radically change a fire's behavior. Modern approaches in fire management incorporate models based on chaos theory to predict how fires spread in varying conditions of terrain and vegetation. This modeling assists in evacuation planning and deployment of firefighting resources more effectively.

In the context of managing these disasters, chaos theory presents a strategic framework that emphasizes flexibility and adaptability in response plans. For example, adjustable evacuation routes and disaster response strategies can be modeled to accommodate various chaotic inputs that might affect the outcome during an actual event. By leveraging chaos-based simulations, planners and responders can prepare for a broader range of scenarios, enhancing resilience against unpredictable natural disasters.

Underpinning all these applications is advanced computational technology paired with extensive data collection from sensors and satellites. As our capability to gather and analyze vast amounts of data increases, so too does our ability to refine our models of chaotic systems, thus progressively enhancing our predictive capacities.

5.7 Long-term Implications for Climate Change

The implications of chaos theory in climate science are profound, allowing researchers and environmentalists to reassess the trajectories of climate change with a new paradigm. Traditional linear models have been crucial but insufficient for understanding the full complexity of the Earth's climate system, which is inherently nonlinear and chaotic. This inability to fully predict the whims of nature due to minute initial uncertainties—termed as the "butterfly effect" in chaos theory—can radically affect long-range climate projections.

At the heart of applying chaos theory to climate science is the utilization of dynamic climate models that integrate chaotic systems to make more accurate predictions. These models take into account small variations in the initial conditions of a complex system, which can lead to vastly different outcomes. This approach helps in understanding how seemingly insignificant changes in the Earth's atmospheric conditions can exponentially amplify and lead to substantial effects on global climate patterns.

Traditional Model	**Chaos-influenced Model**
Assumes smooth progression of climate factors	Incorporates abrupt and discontinuous changes
Predictions more streamlined	Predictions account for larger variability
Based on fixed and stable systems	Adaptative, considering dynamic changes

Consider the recent incidents of polar vortex disruption, which causes sudden and extreme cold weather in temperate regions. These events can be better understood with chaotic models that consider the nonlinear dynamics at play in atmospheric circulation patterns. Without incorporating chaos theory into our understanding, such anomalies could remain largely inexplicable and unpredicted.

In terms of long-term climate change implications, chaos theory highlights that even small interventions or alterations in environmental policy or climate conditions

5.7. LONG-TERM IMPLICATIONS FOR CLIMATE CHANGE

could lead to disproportionately large outcomes, either positive or negative. This suggests an approach where strategic, small-scale interventions may become pivotal in climate change mitigation efforts, challenging traditional broader legislative moves.

The study of phenomena such as El Niño, using chaos theory, reveals insights into how these complex weather patterns may be forecasted more accurately over longer intervals, pertinent for agricultural planning, disaster preparedness, and water resource management. Moreover, chaotic dynamics also play a crucial role in understanding uncertainties associated with climate feedback mechanisms, like cloud formation and ice melt, which are critical elements for refining climate predictions.

Understanding these points helps us to appreciate the necessary changes in how we predict, respond to, and potentially mitigate climate change. Not only does it make forecasting more sophisticated, but it also amplifies our comprehension of how delicate our intervention needs to be. Utilizing chaos theory goes beyond mere prediction—it encourages the continued adaptation and responsiveness of our strategies to combat the impacts of climate change.

Through this lens, educators, policymakers, and climate scientists are urged to cultivate a keen sensitivity to the chaotic intricacies of climate systems. Adopting such sophisticated approaches in modeling can lead to significant inroads in our fight against global warming and climate change, thereby securing a more sustainable future. This refined understanding underscores the critical need for innovative and adaptable strategies that are as dynamic and complex as the systems they aim to forecast and manage.

CHAPTER 5. CHAOS IN NATURE: WEATHER, ECOLOGY, AND BEYOND

Chapter 6

Chaos in Technology: From Internet Traffic to Stock Markets

This chapter examines how chaos theory applies to various technological fields, notably in managing internet traffic and predicting stock market fluctuations. It explores the underlying chaotic behaviors that influence these systems and discusses the methodologies used to model and understand such complexities. Through detailed analysis, the chapter illustrates the impact of chaos theory on improving the robustness, efficiency, and predictability of technology-driven systems, providing key insights for future innovations.

CHAPTER 6. CHAOS IN TECHNOLOGY: FROM INTERNET TRAFFIC TO STOCK MARKETS

6.1 Overview of Chaos in Technological Systems

Chaos theory, a field that studies the behavior of dynamical systems that are highly sensitive to initial conditions—a phenomenon popularly referred to as the butterfly effect—has profound implications in the realm of technology. This section delves into how chaotic dynamics are not just present, but fundamentally influential in technological applications, ranging from the optimization of communication networks to the adaptive algorithms driving the stock market analytics.

The essence of chaos in technological systems can be traced back to the inherent unpredictability and non-linear behavior found in these systems. When we say a system is chaotic, it implies that tiny changes in the starting conditions of the system can lead to vastly different outcomes, making long-term predictions almost impossible with just a superficial knowledge of the initial conditions. This characteristic can be daunting, yet it is also what allows for sophisticated technological advancements through the modeling and understanding of chaotic systems.

To visualize how chaos theory is applied in technology, consider the case of internet traffic. Internet traffic exhibits inherently chaotic characteristics due to the unpredictable and highly dynamic nature of data flows and user interactions. Each packet of data travels through a network influenced by a multitude of factors including network topology changes, fluctuating traffic loads, and varying routing protocols. Even minor perturbations, such as a single router failing or a sudden surge in video streaming due to an unplanned event, can cause signifi-

6.1. OVERVIEW OF CHAOS IN TECHNOLOGICAL SYSTEMS

cant effects on traffic flow, leading to congestion or data loss.

A deeper understanding of chaos in such systems aids in designing more resilient and efficient communication networks. By employing chaotic models, engineers can develop adaptive routing algorithms that predict potential points of failure or congestion, diverting traffic dynamically and minimizing network downtime. These models are built using various chaos-theoretic approaches like Lyapunov exponents to measure the sensitivity of the network to initial conditions or Poincaré maps to observe the states of the network.

In the stock market, chaos theory helps in creating models that adapt to the highly irregular and unpredictable behavior of market prices. Financial markets are complex systems affected by a myriad range of factors including economic indicators, political events, and trader psychology. Traditional predictive models often fall short due to their inability to encapsulate this multifaceted environment. However, by applying chaos theory, more nuanced models can parse through this complexity and offer better predictive power. For instance, techniques such as fractal analysis and nonlinear dynamic modeling allow for a finer understanding of market dynamics, potentially leading to more robust trading strategies that can anticipate drastic market changes.

In drawing these scenarios from communication networks and stock markets, the application of chaos theory in technology appears as both a necessary tool and a lens through which the unpredictability intrinsic to these systems can be understood and mitigated. It fosters innovations that not only cope with the complex characteristics of technological systems but harness them to create smarter, more adaptive technologies.

This integration of chaos theory into technological applications undoubtedly paints a larger picture—a testament to how understanding chaos might lead directly to the very order needed for advancing technological frontiers, turning unpredictable systems into opportunities for innovation and enhanced system robustness.

6.2 Chaos in Communication Networks

Chaos theory, when applied to communication networks, reveals both challenges and opportunities for achieving greater stability and efficiency. Understanding the chaotic nature inherent in these systems is essential for developing protocols that can manage unexpected fluctuations and maintain robust communication channels.

Communication networks, which include cellular networks, the internet, and satellite communications systems, are complex, dynamic systems with inherent nonlinear behaviors. The primary source of complexity arises from the sheer number of interactions between users, devices, and network nodes. Each node in a network can represent a source of unpredictable behavior, influenced by the volume of data traffic, the algorithms used for routing and resource allocation, and the operational environment.

A key aspect of chaos in communication networks is the phenomenon of traffic congestion. Traffic in networks can exhibit sudden changes from fluid flow to congestion due to a variety of factors such as the network design,

6.2. CHAOS IN COMMUNICATION NETWORKS

user behavior, and even minor disruptions like a single failed router. This behavior is analogous to the concept of *bifurcations* in chaos theory, where small changes in the parameters of a system cause a sudden shift in its behavior.

To illustrate this concept, consider a simplified model of a network with nodes connected in a grid-like structure. Each node represents a router, and each connecting line a communication link. The capacity of each link is fixed, but the traffic demand may vary dynamically. A sudden increase in data flow can lead to a congestion collapse, a situation where the total throughput of the network drastically decreases due to severe congestion.

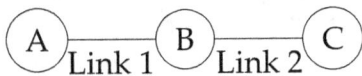

In the above simple network model, if 'Link 1' becomes congested, it not only affects the flow between A and B but also impacts the traffic intended for C via B, demonstrating the interconnected nature of chaos in these systems.

Another significant area within chaotic communications is the modeling and prediction of traffic patterns. Using nonlinear dynamics approaches, researchers have developed models that attempt to predict traffic flows by understanding their chaotic characteristics. These models often employ techniques such as time-series analysis and neural networks to forecast future states based on historical data.

Furthermore, chaos theory provides insights into secure communications. The inherently unpredictable nature of chaotic systems makes them suitable for applications in cryptography. For instance, chaotic maps can generate pseudo-random sequences that can be used for encrypting data transmissions across the network.

One real-world application of chaos theory in communication networks involves adaptive routing algorithms. These algorithms utilize continuous feedback on network conditions to dynamically adjust routes and avoid congestion. By understanding the chaotic patterns of network traffic, these algorithms optimize data paths in real time, mitigating potential bottlenecks and significantly improving the efficiency of data transmission.

By exploring the chaotic elements within communication networks, researchers and engineers can enhance network design and operation, embracing unpredictability to create more adaptable and resilient networks. Through the lens of chaos theory, the next generation of communication networks could effectively manage the dynamic and unpredictable environment of modern data transmission, ensuring reliable and efficient communication in an ever-connected world.

6.3 Internet Traffic and Nonlinear Dynamics

Understanding internet traffic through the lens of chaos theory involves peeling back the layers of daily data transfer to reveal a system governed by complex, nonlinear dynamics. Unlike many physical systems that can be described by linear equations, internet traffic exhibits unpredictable and highly sensitive behavior synonymous with chaotic systems.

6.3. INTERNET TRAFFIC AND NONLINEAR DYNAMICS

At the heart of internet traffic analysis is the concept of self-similarity, a characteristic typical of fractal geometry, where certain patterns repeat at every scale. Studies have shown that internet data packets - whether considered during peak hours or off-peak periods - display a fractal-like structure that does not smooth out with increased traffic but rather becomes increasingly complex. This critical insight forces us to reevaluate traditional Poisson models of internet traffic which assume a certain uniformity and independence in packet arrival times.

The real challenge begins when we apply the principles of nonlinear dynamics to model this behavior scientifically. One way to approach this is through the application of high-dimensional deterministic models, which take into account various variables such as network topology, the protocols in operation, and the types of data being transferred. These models are designed to capture the inherent unpredictability and the potential for sudden shifts in traffic flow, fundamentally attributing these phenomena to initial conditions at infinitesimal scales.

One practical application of understanding this chaotic behavior has been in the development of adaptive algorithms capable of adjusting the flow of data dynamically. Consider, for instance, the algorithm that modulates data packet transmission based on real-time analysis of network congestion. By continuously adjusting itself, the algorithm optimizes throughput while preventing overload conditions, a crucial feature for maintaining service quality in high-demand scenarios.

Further insights can be visually represented through phase space diagrams, where the changes in internet traffic can be modeled over time. These diagrams, typically employed in the study of dynamical systems, illustrate trajectories or orbits of the states of a system based on a set of initial conditions. For internet traffic, these states can be numerical representations of various metrics such as packet loss rates, number of concurrent connections, or transmission times, among others. A simplified version of such a diagram could be plotted as follows:

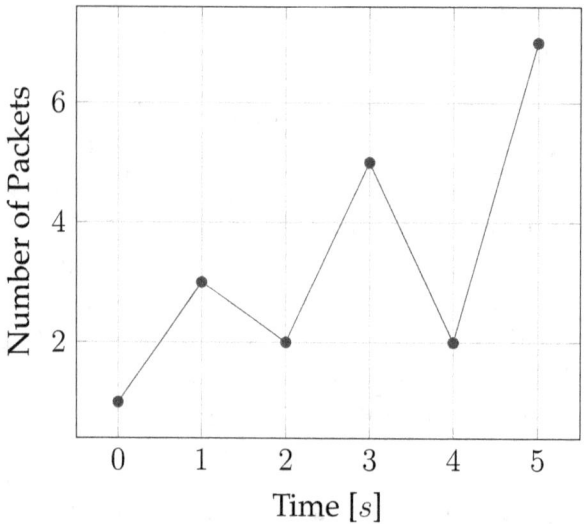

Phase Space Diagram for Internet Traffic

As we delve deeper into the chaotic characteristics of internet traffic using nonlinear dynamics, we pave the way not only toward more robust network designs but also towards a more fundamental understanding of how unpredictability can be harnessed to optimize technological systems. This approach brings out a paradoxical element: by embracing the chaos inherent in internet traffic, we can achieve higher levels of control and predictability in network performance.

6.4 Stock Market Fluctuations and Predictive Models

The arena of financial markets exemplifies a typical chaotic system where minute changes in initial conditions can significantly impact outcomes. Stock markets, with their inherently complex behaviors, provide a ripe field for applying chaos theory to understand and perhaps forecast future movements. By dissecting the intricate dynamics of these fluctuations through a chaotic lens, one can gain insights into the seemingly random but deeply deterministic patterns that drive market behaviors.

Stock market data are influenced by a multitude of factors, including economic indicators, political events, investor psychology, and global incidents, among others. Due to the interconnectivity and interdependence of these factors, the financial markets exhibit non-linear dynamics. Such systems are sensitive to initial conditions — a hallmark of chaotic systems — which makes precise long-term predictions challenging yet not entirely impossible.

To model these dynamics, economists and mathematicians have developed various predictive models that incorporate elements of chaos theory. Traditional models like the ARIMA (AutoRegressive Integrated Moving Average) have been expanded upon to include nonlinear components and elements of chaos theory, leading to improvements in predictive capabilities. More recently, techniques involving machine learning and artificial intelligence have entered the fray, utilizing large datasets and the ability to learn from previous market conditions to predict future trends.

Phase Space Reconstruction is a critical technique employed in understanding stock markets through chaos theory. This method helps in visualizing the trajectory of stocks in a multidimensional space, providing insights into the cycles and patterns that prevail over time. By reconstructing phase space, analysts can identify chaotic attractors that guide the system's behavior, allowing for more perfect insights into the market's movements.

Another aspect vital to chaos in stock markets is *fractal analysis*. Markets display fractal behavior; they are self-similar across different timescales. This revelation has profound implications for the predictability and analysis of financial markets. By applying fractal mathematics — quantified through methods like the Hurst exponent — analysts can better understand market conditions and volatility, predicting periods of stability or fluctuation with greater accuracy.

The application of **Neural Networks** has also been transformative. These networks mimic human brain function to an extent and learn from large amounts of data, identifying patterns that are not immediately obvious. By training neural networks on historical stock market data — incorporating fundamentals like corporate earnings, geopolitical impacts, and macroeconomic variables — better nonlinear forecasting models have been developed. These models help in understanding the underlying chaos and making informed predictions about future market directions.

In considering the practical application of these tools and models, one must also address the ethical implications. The capability to predict stock market trends can lead to significant financial gains but also raises concerns about market manipulation and fairness. Therefore, while chaos theory offers substantial benefits in understanding and forecasting market behaviors, it necessitates responsible usage to ensure market integrity and investor protection.

By integrating chaos theory into predictive models, the approach to dealing with stock market fluctuations has radically changed. These models furnish finance professionals with more robust tools to anticipate market changes, albeit recognizing the inherent limitations imposed by chaotic systems. This innovation not only aids in better comprehending the complexities of the financial world but also sets the stage for future advancements that could continue to revolutionize our approach to economic forecasts.

Through the lens of chaos theory, we are witnessing a shift towards more sophisticated and nuanced modeling techniques in financial markets, marking a significant move from merely reactive strategies to more proactive and informed ones. This transition underscores an ever-growing appreciation for chaos as not merely a challenge to overcome but as a profound opportunity to harness.

6.5 Chaos in Electronic Circuits and Devices

Chaos theory, while predominantly explored in natural sciences and economics, also finds intriguing applications within the realm of electronics. Specifically, the unexpected behavior of chaotic dynamics is observed in electronic circuits and devices, which, despite appearing as a negative trait, can be harnessed for innovative applications such as in secure communications and novel computing methods.

An electronic chaos generator is a fundamental building block for studying chaos in electronics. These are typically simple electronic circuits that use nonlinear components like diodes, operational amplifiers, or nonlinear resistors to create their chaotic outputs. The chaotic behavior in these circuits is highly sensitive to initial conditions as well as to the intrinsic properties of the system components.

Understanding Chaotic Circuits One commonly studied chaotic circuit is the Chua's circuit. It consists of two capacitors, an inductor, a linear resistor, and a nonlinear resistor called the Chua's diode. The circuit displays a rich variety of dynamic behaviors, including periodic oscillations and chaos. The mathematical model describing the Chua's circuit includes a set of three nonlinear differential equations, where the nonlinearity stems from the voltage-current characteristic of Chua's diode.

To illustrate, consider the dynamics of the Chua's circuit represented by:

$$\begin{cases} C_1 \frac{dv_{C1}}{dt} = \frac{1}{R}(v_C 2 - v_C 1) - i_{nl}(v_C 1), \\ C_2 \frac{dv_{C2}}{dt} = \frac{1}{R}(v_C 1 - v_C 2) + i_L, \\ L \frac{di_L}{dt} = -v_C 2. \end{cases}$$

Here, $i_{nl}(v_{C1})$ represents the current through the nonlinear resistor, which adds the essential nonlinearity leading to chaos.

Application in Secure Communications One of the most promising applications of chaotic circuits is in secure communications. The idea is to use chaotic signals

6.5. CHAOS IN ELECTRONIC CIRCUITS AND DEVICES

to mask the transmission of information. Due to the sensitive dependence on initial conditions, even a slight difference in starting conditions between a transmitter and a receiver can lead to significantly different chaotic outputs, which makes these systems inherently secure against attackers without access to the exact system parameters and states.

A common method is to generate a chaotic carrier in both transmitter and receiver circuits. The information signal modulates the chaotic carrier before transmission. The receiver, exploiting the synchronization properties of chaotic systems, can recover the modulated signal. This method effectively uses chaos as a means to encrypt and decrypt signals, thereby enhancing security in data transmission.

Chaos-Based Computing Beyond communications, chaotic circuits find applications in emerging computing paradigms like neural networks and optimization algorithms. Chaotic dynamics are used to enhance randomness and prevent local minima problems in algorithms like simulated annealing or genetic algorithms. For instance, chaotic sequences can replace traditional pseudorandom number generators to provide higher randomness and improve the robustness of stochastic processes.

With advances in technology and deeper understanding of chaos theory, electronic circuits that exhibit chaotic behavior offer a fertile ground for research and development in enhancing electronic device functionality. The adoption of chaotic principles in circuits and devices not only challenges traditional designs but also paves the way for developing new technologies that leverage the unpredictability and complexity of chaotic systems for advanced technological applications.

6.6 Applications in Cryptography and Security

The advent of chaos theory has significantly impacted various fields, with cryptography and security standing out as primary beneficiaries. This synergy arises from the foundational nuances of chaos theory—sensitivity to initial conditions, unpredictability, and the ability to resemble random behavior. These characteristics align impeccably with the fundamental requirements for cryptographic systems: secrecy and robustness to attacks.

Chaos-based algorithms translate these chaotic properties into cryptographic techniques. One salient aspect is the implementation of chaos in encryption methods. A typical approach employs a chaotic map, often a simple mathematical function exhibiting chaotic behavior over time. These functions are leveraged to generate pseudorandom sequences which, due to their complexity and sensitivity to initial values, serve as keys in encryption algorithms.

Let's delve into a practical example: Consider the logistic map, defined by the equation $x_{n+1} = rx_n(1-x_n)$, where r is a parameter influencing the system's behavior. By carefully choosing r, the sequence becomes chaotic, producing a pseudorandom series hardly distinguishable from true random sequences at a quick glance. Encryption algorithms can use this series as a cryptographic key, encoding messages in such a way that without the exact initial conditions and parameter r, decoding the encrypted

6.6. APPLICATIONS IN CRYPTOGRAPHY AND SECURITY

message becomes computationally impracticable.

Further extending this application is the use of chaotic maps in generating digital signatures. Digital signatures ensure the authenticity and integrity of a message or document in digital communication. By integrating chaotic maps, the sensitivity to initial conditions of chaos ensures that even a minimal change in the document results in a substantially different signature, thereby safeguarding against unauthorized modifications and forgeries.

Another intriguing aspect is the implementation of chaos in secure communication channels, particularly relevant in military and financial communication systems. Here, chaos synchronizes two systems at either end of the transmission channel. Each system operates in a chaotic regime, and synchronization occurs so subtly that intercepting the communication without detection becomes exceedingly difficult. This method offers a high-security margin as any slight deviation in the system parameters disrupts the synchronization, making unauthorized access ineffective.

Visualizing Chaos in Cryptography with Pseudocode
To better understand how a chaotic encryption process might be implemented, consider the following simplified pseudocode using the logistic map for generating a key:

```
Initialize: x0, r
For i from 1 to N do
    x[i] = r * x[i-1] * (1 - x[i-1])
End for
Output: Sequence x[1] to x[N] as key
```

This pseudocode snippet highlights the iterative nature and sensitivity to the initial conditions, both cardinal traits of chaotic systems used in cryptography.

By capitalizing on these chaotic properties, cryptography not only enhances the security levels but also adds a layer of dynamic defense mechanisms that are more adaptable to evolving threats in digital communications environments. As technology progresses, the integration of chaos theory in security strategies will invariably become more pronounced, further strengthening the wall against cyber threats and unauthorized data breaches.

6.7 Adaptive Technologies Harnessing Chaos

The integration of chaos theory into adaptive technologies is a frontier in modern engineering that promises to enhance system responsiveness and efficiency in unprecedented ways. Adaptive technologies, which adjust their operation based on changes in their environment or input, are natural candidates for the application of chaotic dynamics to improve their performance.

Central to the discussion of adaptive technology is the concept of chaotic optimization. This technique employs chaotic maps to generate a diverse set of potential solutions to optimization problems, thereby avoiding the pitfalls of conventional methods which may fall into local minima. Chaotic optimization exploits the inherent irregularities in chaotic systems to explore the solution space more thoroughly. Applications are vast, ranging from robotic path planning where chaos aids in navigating complex, unforeseen terrains, to dynamic resource allocation in computer networks.

Moreover, the application of chaos theory is profoundly evident in the field of artificial neural networks (ANNs), particularly in systems that require real-time learning and adaptation. ANNs benefit from chaotic sequences which prevent the convergence to sub-optimal solutions and encourage exploration of the parameter space. This is crucial in environments where the input data may exhibit non-linear or unpredictable patterns, thus necessitating a flexible and adaptive response.

Schematic of an ANN with chaotic sequence integration

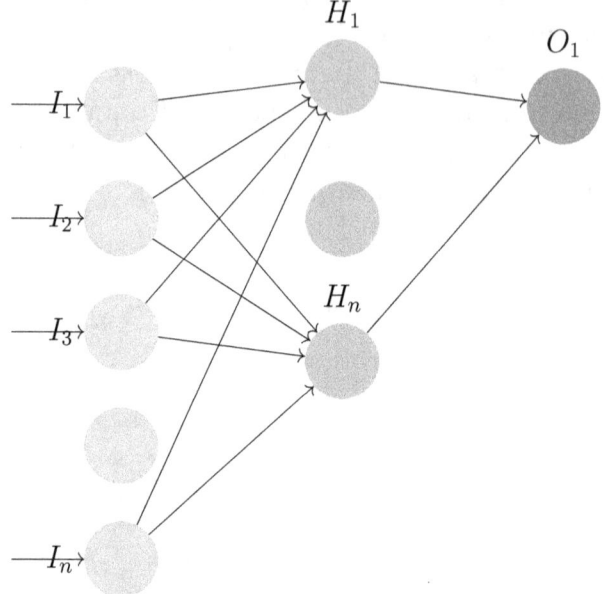

The synergistic effect of chaotic dynamics in adaptive algorithms has also been applied to stochastic optimization techniques such as genetic algorithms and swarm intelligence. By incorporating chaotic variables into these algorithms, researchers have observed improvements in convergence rates and solution quality. The erratic nature of chaotic sequences introduces a beneficial component of randomness that helps explore more robust solutions quickly.

Another exciting development in harnessing chaos is seen in adaptive signal processing used in telecommunications. Here, chaotic modulation techniques are employed to enhance the security and reliability of data transmission. The unpredictable nature of chaotic signals makes them excellent for securing communication from interceptive threats, while also improving the fidelity of signal transmission over noisy channels.

The future direction of adaptive technologies influenced by chaos theory leans heavily towards developing smarter, more efficient systems capable of learning and adapting from past behaviors without human intervention. As we continue to imbue machines with the capability to harness chaos effectively, the boundaries of technology's capability will expand, leading to smarter and more adaptable systems. This progression not only highlights the complexity and unpredictability of technological systems but also showcases the potential for chaotic principles to bring significant advancements.

The paradox of finding order in chaos reflects in how these technologies learn to anticipate and adapt to the seemingly random fluctuations in their operating environments. This mirrors the broader scope of chaos theory — harnessing complexity to foster advancement and innovation.

6.8 Challenges and Future Directions in Technological Chaos

One of the most pressing challenges in employing chaos theory within technological frameworks is the reliable prediction and control of chaotic systems. Due to their inherent sensitivity to initial conditions, small errors in measurement or in the parameters used in predictive models can lead to vastly different outcomes. This sensitivity, often referred to as the "butterfly effect," implies that even with advanced computational techniques, long-term predictions remain precarious.

The difficulty of modeling chaotic systems accurately poses another significant challenge. Traditional linear models fail to capture the dynamic complexities exhibited by chaotic systems. As a result, there is a continuous need for the development of innovative nonlinear modeling techniques that can better represent the intricate behaviors of these systems. The integration of machine learning with chaos theory presents a promising approach, offering the potential to improve the accuracy of predictions by learning from vast amounts of data and identifying hidden patterns within chaotic dynamics.

Furthermore, the practical implementation of chaotic systems in technology also encounters hurdles. In areas such as secure communication and cryptography, while chaos provides enhanced security features due to its unpredictability, it also requires precise synchronization between the sender and receiver. Any slight deviation in synchronization can lead to complete failure in communication, thus raising the importance of developing robust methods for maintaining synchronization in chaotic systems.

As we look to the future, one exciting direction is the exploration of quantum chaos and its implications for technology, particularly in quantum computing and secure communications. Quantum systems exhibit chaotic behavior under certain conditions, and understanding these dynamics could be key to developing more powerful quantum algorithms and tamper-proof security systems.

Another promising direction involves the use of chaos theory in enhancing artificial intelligence systems. By embedding chaotic models in neural networks, there is the potential to create AI systems that can better mimic human cognitive processes, which are inherently nonlinear and possibly chaotic. This integration could lead to significant advancements in how AI systems learn and adapt, pushing the boundaries of current technologies.

To harness the full potential of chaos in technology, interdisciplinary collaboration will be essential. Physicists, mathematicians, computer scientists, and engineers must work together to overcome the barriers posed by chaotic systems. This collaborative effort will not only deepen our understanding of chaos but also expand its applications, contributing to technological innovations that were previously considered unattainable.

Exploring chaos theory in technology thus opens up a fascinating frontier filled with both formidable challenges and incredible opportunities. Leveraging this theory effectively will enable us to devise more robust, efficient, and innovative technological solutions, shaping the future of technology in unforeseen ways.

CHAPTER 6. CHAOS IN TECHNOLOGY: FROM INTERNET TRAFFIC TO STOCK MARKETS

Chapter 7

Predicting the Unpredictable: Limitations and Possibilities

This chapter delves into the challenges and methodologies associated with predicting outcomes in chaotic systems, where traditional forecasting methods often fail. It discusses the theoretical limits of predictability and explores advanced techniques such as machine learning and dynamic modeling, which offer new possibilities for forecasting in the face of inherent unpredictability. By examining both the potentials and limitations, the chapter provides a balanced view on the current state and future prospects of prediction in fields impacted by chaos theory.

CHAPTER 7. PREDICTING THE UNPREDICTABLE: LIMITATIONS AND POSSIBILITIES

7.1 The Paradox of Predicting Chaos

Deep within the study of chaos lies a fascinating contradiction: although chaotic systems are deterministic, meaning their future dynamics are fully determined by their initial conditions, they also exhibit extreme sensitivity to these conditions, leading to what is commonly known as the butterfly effect. This inherent sensitivity makes long-term predictions highly challenging, yet not altogether impossible.

Chaos theory teaches us that small differences in initial conditions can yield vastly different outcomes. This property, known as sensitive dependence, is typified by the famous metaphor where a butterfly flapping its wings in Brazil can ultimately cause a tornado in Texas. Mathematically, this is often represented by Lyapunov exponents, which measure the rate at which nearby trajectories diverge in a dynamical system. For chaotic systems, positive Lyapunov exponents are indicative, suggesting that minor changes will grow exponentially over time.

This notion poses a significant challenge for prediction, considering that it is practically impossible to measure initial conditions with absolute precision. Any small error in measurement can lead to widely diverging outcomes, rendering long-term forecasts highly speculative. Yet, it is this very unpredictability that often governs the behavior of many natural and man-made systems, from the weather patterns influenced by climatic chaos to the fluctuating stock markets driven by economic dynamic systems.

Despite these challenges, the concept of predictability in chaos is not a lost cause but instead a nuanced problem. In shorter timescales, chaotic systems can exhibit transient predictability due to their deterministic nature. During these periods, trajectories remain sufficiently close to allow for some level of accurate forecasting before diverging paths become pronounced. This window of predictability, although limited, can be extremely valuable in various practical applications such as weather forecasting, where even a short lead time can be critical.

Moreover, advancements in computational power and techniques have opened new avenues for managing and predicting chaotic systems. Numerical methods for solving differential equations that describe chaotic systems have improved, allowing for better simulations of their behavior under different initial conditions. This progression offers a glimmer of hope that with increasing computational resources and sophisticated algorithms, the horizon of effective prediction can be extended further.

Ticks characterized by penetrating episodes of predictability within chaotic systems invite curiosity and deeper inquiry. It leads to rich insights into how chaos and order are not always opposites but can coexist in a complimentary and often surprisingly coherent fashion. Despite its daunting appearance, the realm of chaos offers a unique perspective on the universe's complexity, encouraging thinkers and researchers to explore beyond traditional boundaries of knowledge and prediction techniques.

Exploring these boundaries allows us to appreciate the delicate balance and dance between knowing and not knowing — between being able to predict what will happen next and stepping into the vast unknown. It invites a deeper examination of how we interact with our world and encourages a humility appropriate to the inherent limits of our knowledge. This interplay between predictability and unpredictability not only challenges our understanding but also enhances our appreciation for the complexity of the systems that shape our world.

7.2 Methods of Forecasting Chaotic Systems

Forecasting chaotic systems represents a considerable challenge due to their inherent sensitivity to initial conditions and seemingly random nature. However, several sophisticated methods have been developed to enhance the predictability of these systems, focusing particularly on short-term forecasting. This section explores three main techniques: nonlinear dynamic modeling, state space reconstruction, and ensemble forecasting.

Nonlinear Dynamic Modeling is extensively used in forecasting chaotic systems and operates under the premise that even chaotic systems follow deterministic rules. By constructing mathematical models that describe these rules, predictions about future states of the system can be generated. The Lorenz system, for example, illustrates how differential equations can model weather patterns despite their chaotic nature. When deploying this method, it is key to accurately capture the dynamics through differential equations that delineate how the state of the system evolves over time. For more precision, these models often incorporate parameters refined by historical data, enabling them to adapt and potentially improve accuracy with new inputs.

State Space Reconstruction is a method derived from Takens' Theorem, which posits that the dynamics of a chaotic system can be reconstructed from a series of obser-

7.2. METHODS OF FORECASTING CHAOTIC SYSTEMS

vations using delay embedding techniques. This involves mapping time series data into a higher-dimensional phase space, where the system's dynamics become more discernible and thus more predictable. The embedding dimension and delay time are critical parameters in this method. They must be carefully selected based on the underlying data to adequately unfold the dynamics of the chaotic system. Visualizations often assist in determining the appropriate parameters by revealing the system's structure in the reconstructed state space.

Ensemble Forecasting tackles the unpredictability inherent in chaotic systems by using multiple forecasts that simulate various possible future states. Each forecast in the ensemble represents a potential future state, considering different initial conditions sampled around the current state's estimate. This method acknowledges the sensitivity to initial conditions by not relying on a single prediction but instead examining a distribution of possible outcomes. The spread and central tendency of these forecasts can provide insights into the most likely scenarios and the associated uncertainties. Ensemble forecasting is particularly prevalent in meteorology, where multiple weather models with slightly varied initial conditions are used to forecast severe weather events.

Each of these methods encompasses a unique approach to mitigating the challenges posed by chaotic systems. Nonlinear dynamic models offer a deterministic foundation, while state space reconstruction focuses on uncovering hidden patterns through geometrical representations. Ensemble forecasting, on the other hand, leverages probabilistic approaches to account for the inherent uncertainties and sensitivities. By integrating these techniques, forecasters can improve their predictions' reliability, albeit within the time limits dictated by the chaotic nature of the systems.

Understanding these methods' subtleties and correctly applying them requires deep knowledge of both the theoretical aspects of chaos theory and practical expertise in statistical and dynamic modeling. Innovations continue to emerge as researchers combine these established techniques with new computational tools and machine learning algorithms, driving forward the precision and applicability of forecasting in chaotic environments.

As we delve deeper into chaos theory's implications across various fields, it becomes evident that mastering these forecasting techniques not only enhances our understanding of chaotic systems but also equips us with better tools to navigate and possibly foresee the complexities of nature's unpredictability.

7.3 Uncertainty and Errors in Predictions

The endeavor to predict outcomes within chaotic systems confronts inherent uncertainties, which stem from numerous sources and manifest as various types of errors. Chief among these is the sensitivity to initial conditions, famously illustrated by the butterfly effect, which suggests that infinitesimal changes in the initial state of a system can result in vastly different outcomes. This section delves deeper into the nature of these uncertainties and the categorization of errors they produce, crucial for refining the predictive models used in chaos theory.

Sensitivity to Initial Conditions: In chaotic systems, the initial conditions determine the long-term behavior, but

7.3. UNCERTAINTY AND ERRORS IN PREDICTIONS

pinpointing these conditions with absolute precision is practically unfeasible. This sensitivity is quantified by the Lyapunov exponent, which measures the rate at which trajectories diverge. For predictions, a positive Lyapunov exponent is significant, implying that the error in any forecast will grow exponentially with time. Even with advanced computational techniques, the amplification of minor inconsistencies in initial data accumulates, thereby escalating predictive uncertainty.

Model Error: Another major contributor to prediction errors in chaotic systems is the inherent imperfections in the models themselves. Every mathematical model is a simplification or approximation of the real world and thus, inherently imperfect. These imperfections might be due to assumptions made for model simplicity or computational constraints. For instance, in weather forecasting, simplifications are necessary in representing complex atmospheric processes. As a result, small-scale phenomena and their interactions with larger systems are occasionally modeled inadequately, leading to discrepancies between predicted and actual outcomes.

Data-driven Challenges: Accurate predictions also hinge upon the quality and quantity of data. In chaotic systems, where the outcome is highly sensitive to initial states, incomplete or noisy data can lead to significant errors. Moreover, in many systems of interest, collecting exhaustive data sets is not always feasible due to technological and logistical limitations. For instance, in ecological systems, monitoring every influencing factor is impractically intensive, introducing gaps in the data which translates directly into predictive imprecision.

Ultimately, managing these uncertainties and minimizing errors necessitates an ongoing refinement of data collection processes, model development, and computational methods. As we forge ahead, the intertwining of advanced machine learning techniques with traditional dynamic modeling provides a promising avenue to enhance our predictive capabilities in the realm of chaos theory. By continuously incorporating new data and adapting to observed outcomes, the iterative refinement of predictions allows for a progressive reduction in uncertainty, paving the way for more reliable forecasts even in inherently unpredictable systems.

7.4 Limitations of Current Models

Despite significant advances in computational power and methodological sophistication, the predictive models utilized in studying chaotic systems still grapple with several inherent limitations. Understanding these limitations not only broadens our comprehension of chaotic dynamics but also guides future improvements in prediction technology.

One fundamental limitation of current forecasting models lies in their dependency on initial conditions. Chaotic systems, by their nature, are highly sensitive to these initial states; a minuscule alteration can result in vastly different outcomes, known within chaos theory as the "butterfly effect." However, measuring these initial conditions with absolute precision is practically impossible due to inherent limitations in the accuracy and resolution of measurement instruments. The exponential growth of error with time in the dynamics of chaotic systems restricts the practical predictability horizon, beyond which predictions become unreliable.

7.4. LIMITATIONS OF CURRENT MODELS

Moreover, most models are constrained by the assumptions and simplifications made during their development. These models often rely on approximations to manage the complexity of chaotic systems or to make the problems computationally tractable. For instance, when physical processes are approximated or important variables are ignored, the model might not capture the full dynamical behavior of the system. This skewing of system representation often leads to a divergence between the model's predictions and actual outcomes.

Another significant limitation is the issue of computational resource requirements. Advanced models that incorporate a high level of detail and more variables tend to require enormous computational power and time, limiting their practical usage, especially for real-time applications. This computational demand constrains the ability of researchers to run multiple simulations with varying initial conditions or parameters, which is essential for exploring the diverse possible outcomes in a chaotic system.

Furthermore, the deterministic models that are frequently used for predicting chaotic dynamics can inherently not accommodate the stochastic nature of real-world systems. Real-world systems often undergo fluctuations due to random external influences which are not accounted for in deterministic models. This missing stochastic component can lead to significant predictive inaccuracies when the modeled system is influenced by random, unpredictable factors.

Lastly, while numerical methods employed in modeling such as finite differences or spectral methods are powerful, they introduce discretization errors. Even with fine discretization, tiny errors can grow exponentially fast due to the sensitive dependence on initial conditions characteristic of chaotic systems. Additionally, numerical schemes might suffer from stability and convergence issues when applied to highly nonlinear equations typical in chaos theory.

To visualize the effect of some of these limitations, consider a hypothetical experiment involving a Lorenz attractor, a classic example of a chaotic system. If we plot trajectories starting from slightly different initial conditions using a standard computational model, we would observe the divergence of these trajectories over time, illustrating the sensitivity to initial conditions and the exponential growth of error in predictions.

By examining these limitations, efforts can shift towards refining current models or creating new methodologies that address these weaknesses, enhancing our ability to forecast chaotic systems with greater accuracy and over longer time horizons. Adaptive modeling techniques and hybrid models incorporating stochastic elements could potentially offer robust frameworks for improving prediction in chaotic systems, leading us towards more reliable and effective forecasting tools.

7.5 Advances in Computational Techniques

The field of chaos theory is rich with complexities, and traditional approaches often struggle to grasp the full nuance of chaotic systems. However, recent advances in computational techniques have provided new tools and methods that can significantly enhance our ability to understand and predict these enigmatic systems.

7.5. ADVANCES IN COMPUTATIONAL TECHNIQUES

One of the standout techniques that has evolved dramatically in recent years is high-resolution simulation. With the advent of supercomputing and parallel processing architectures, scientists can now perform simulations at scales and speeds that were once thought impossible. These advancements allow for the detailed modeling of chaotic systems across various fields, from meteorology to engineering. For example, in weather forecasting, enhanced computational capabilities enable the integration of massive data sets into dynamic models, offering predictions that are both more accurate and granular.

Another significant advancement is the development of algorithmic improvements in numerical methods that specifically address the non-linear dynamics characteristic of chaotic systems. Traditional numerical methods often fall short when applied to these systems due to their sensitivity to initial conditions and potential for exponential divergence. New algorithms, however, employ adaptive techniques that modify the computation step sizes dynamically, thereby improving the accuracy of simulations without a proportional increase in computational expense.

The use of ensemble methods also represents a crucial advancement in the computational analysis of chaotic systems. Rather than relying on a single model or prediction, ensemble methods involve running multiple simulations with slightly varied initial conditions or model parameters. This approach not only provides a spectrum of possible outcomes but also quantifies the uncertainty inherent in predictions of chaotic systems. By examining the spread and general tendencies of the ensemble's results, researchers can gain a more reliable insight into the most probable outcomes while accounting for the system's inherent unpredictability.

Transitioning further towards cutting-edge technology, quantum computing has begun to make its mark. Although still in its nascent stages, the potential for quantum computing to process information at previously unattainable speeds poses a promising horizon for chaos theory. The peculiarities of quantum mechanics could allow for the simulation of chaotic systems in ways that classical computers cannot, possibly leading to breakthroughs in predictive accuracy and speed.

Lastly, the integration of machine learning with traditional computational methods forms a frontier of research that melds data-driven insights with dynamic modeling capabilities. Machine learning models, particularly those utilizing deep learning architectures, are adept at identifying patterns and predicting outcomes from large datasets, which are often noisy and unstructured. When applied to chaotic systems, these models can uncover underlying patterns that are not immediately apparent through conventional analysis.

These advancements all contribute to enhancing our profile of chaotic phenomena, bridging gaps between theoretical understanding and practical application. With these tools at our disposal, we edge closer to not just predicting but also controlling aspects of the chaotic systems that permeate our world, a vision that, while still distant, is increasingly within our grasp.

7.6 Machine Learning and AI in Prediction

The integration of machine learning (ML) and artificial intelligence (AI) technologies into the study of chaotic sys-

7.6. MACHINE LEARNING AND AI IN PREDICTION

tems has ushered in a transformative era in predicting the unpredictable. By leveraging complex algorithms, these technologies can discern patterns and make predictions from massive, noisy, and dynamic datasets where traditional methods falter.

Machine learning models, particularly deep learning networks, are adept at processing and analyzing extensive data arrays to identify hidden patterns that precede chaotic outbursts. The crux of ML lies in its ability to continually learn from data without being explicitly programmed to do so. This feature is crucial in environments governed by chaotic dynamics, where slight changes in initial conditions can lead to drastically different outcomes.

A typical application is the use of recurrent neural networks (RNNs), especially Long Short-Term Memory (LSTM) networks, which have shown impressive results in sequence prediction problems. These models are capable of learning from sequences of data, making them ideal for time-series forecasting, a common scenario in chaotic systems like weather patterns and stock markets.

In addition to neural networks, reinforcement learning (RL) has been applied to chaotic systems to great effect. In RL scenarios, algorithms learn optimal strategies through trial and error, interacting with a dynamic environment in which the conditions are continually changing. This method mirrors the continuous adaptation needed to manage and predict within chaotic systems effectively.

However, the integration of ML and AI also presents several challenges. One of the primary concerns is the "black box" nature of many deep learning models, where the decision-making process is not transparent or easily understandable. This opacity can be problematic in fields where prediction requires not just accuracy, but also reliability and trustworthiness, like in autonomous driving or medical diagnostics.

To address these challenges, researchers are exploring explainable AI (XAI) methods that aim to make the predictions of AI systems more interpretable and trustworthy. Techniques such as feature importance visualization and model-agnostic methods are among the tools being used to peel back layers of ML models, providing insights into their predictive processes.

Here are some notable results from recent studies:

- An LSTM network trained on historical hurricane tracks successfully predicted subsequent paths with greater accuracy than traditional meteorological models.

- A study utilizing convolutional neural networks (CNNs) for predicting the onset of chaos in fluid dynamics outperformed legacy computational fluid dynamics models.

- Reinforcement learning techniques were employed to optimize operations in a factory setting where machinery conditions dynamically change, preventing recurrent breakdowns and enhancing efficiency.

To illustrate the effectiveness of these approaches, consider this simplified model of a chaotic system predicted by an LSTM network:

```
Model: Sequential
    Layer (type)                 Output Shape              Param #
=================================================================
    lstm_1 (LSTM)                (None, 100)               40400
```

```
dropout_1 (Dropout)        (None, 100)              0
dense_1 (Dense)            (None, 1)              101
=================================================================
```

This example underscores how layers and neurons configure to capture and use historical data to predict future states in a linear output format representing the predicted condition.

While challenges persist, the potential of ML and AI to revolutionize predictions in chaotic systems remains substantial. Researchers continue to refine these models, enhancing their capacities to deal with the inherent unpredictabilities of chaotic environments. The future landscape of chaos prediction, bolstered by AI and ML innovations, holds exciting possibilities that promise not only more refined predictions but also novel insights into chaotic dynamics that were previously thought to be beyond reach.

7.7 Real-world Case Studies

Let us explore several real-world applications and case studies where chaos theory has been successfully applied to model and predict outcomes, demonstrating both the potential and the challenges inherent in dealing with chaotic systems.

One of the most well-known applications of chaos theory is in meteorology. Weather systems are inherently unpredictable due to their chaotic nature, but advancements in dynamic modeling have improved short-term weather forecasting significantly. For instance, the Lorenz system, a set of deterministic non-linear differential equations, was one of the first examples to emphasize the chaotic nature of weather systems. Even with state-of-the-art technology, the accuracy of weather forecasts significantly drops beyond a few days. Detailed analysis of these models, involving continuous data assimilation and correction using real-time observational data, underpins modern meteorological predictions.

Another fascinating case study is the application of chaos theory to population biology, specifically in predicting fluctuations in wildlife populations. The Canadian Lynx and Snowshoe Hare pelt records from the Hudson Bay Company over nearly a century depict a clear 10-year cycle. Such population dynamics can be modeled using differential equations that account for both prey and predator interactions, displaying chaotic solutions under certain parameters. These models help in understanding complex ecological interactions and are vital for conservation planning and resource management.

Chaotic behavior also extends to the realm of cardiology, particularly in understanding and predicting heart rhythms. Heartbeat intervals exhibit nonlinear dynamics that can become chaotic, which is often associated with pathological conditions such as cardiac arrhythmias. Using techniques derived from chaos theory, researchers can analyze electrocardiogram data to predict sudden cardiac arrests even when traditional methods fail to indicate such risks. This approach provides a potentially lifesaving analysis by identifying subtle changes in heartbeat dynamics, offering a chance for preventative action.

The stock market is another area notoriously known for its unpredictable behavior, often described as chaotic. Financial economists utilize chaos theory to model and predict the stock market movements. Through dynamic modeling of stock prices and trading volumes, patterns that appear random may reveal a chaotic structure that can be analyzed and, to some extent, predicted. This understanding can significantly affect investment strategies and risk assessment models in the finance industry.

Visualization and Analysis:

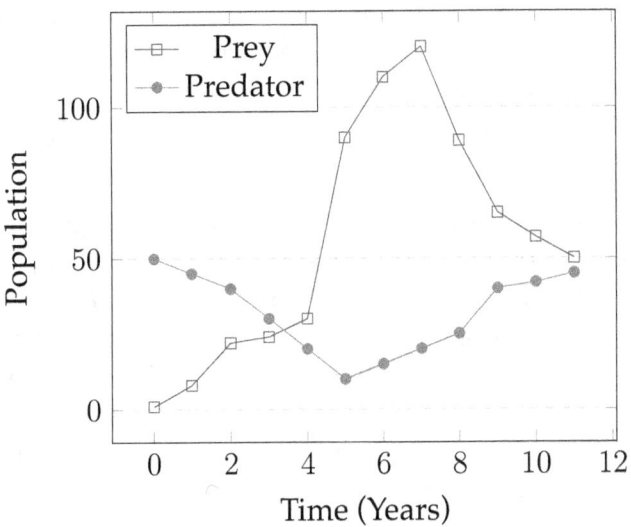

Through these diverse examples spanning meteorology, biology, cardiology, and finance, it becomes evident that understanding and applying chaos theory can lead to better predictive mechanisms. Despite the inherent unpredictability dictated by chaos, strategic applications of chaotic models allow for a deeper insight into complex systems. This approach not only enhances forecasting accuracy but also aids in strategic planning across various scientific and commercial fields, underscoring the critical role of chaos theory in tackling predictably unpredictable phenomena.

7.8 Ethical Considerations and Responsible Forecasting

Predictive modelling in chaotic systems, while opening new vistas of forecasting, necessitates a conscientious evaluation of the ethical implications associated with its application. As these techniques become increasingly central in sectors such as finance, healthcare, and public safety, the responsibility to deploy these technologies judiciously becomes paramount.

Firstly, it's crucial to acknowledge the inherent biases that can be embedded in predictive models. Machine learning algorithms, a cornerstone of modern predictive analytics, often learn from historical data. If this data contains biases, the predictions will likely perpetuate and even amplify these biases, leading to skewed or unfair outcomes. For instance, in healthcare, a model trained on data from a population that does not adequately represent all demographic groups may result in less effective or inappropriate medical interventions for underrepresented groups. Ethical forecasting must thus involve rigorous scrutiny of training data and continual reassessment of model outcomes to ensure fairness and inclusivity.

Transparency represents another crucial ethical pillar. Stakeholders, particularly those potentially affected by predictions, deserve to understand how predictions are made and the degree of uncertainty involved. This transparency is essential not only for fostering trust but also for ensuring that users of predictive models maintain realistic expectations about their capabilities and limitations. In financial markets, for example, where predictive models can influence significant economic decisions, stakeholders must be explicitly informed about the potential risk of dependency on predictions derived from inherently unpredictable chaotic systems.

The ethical use of predictive models in chaotic systems further extends to privacy concerns. As these models often require massive amounts of data, including personal information, ensuring privacy and security against breaches becomes a non-negotiable requirement. Moreover, as data is increasingly cross-referenced and integrated from diverse sources to enhance predictive accuracy, the potential for invasive surveillance and loss of individual privacy escalates. Ethical forecasting must, therefore, place stringent controls on data use, uphold robust security protocols, and prioritize individual privacy rights without compromising the effectiveness of the predictions.

Accountability in predictive modeling is intricately linked to its ethical application. When predictions fail— as they inevitably do in chaotic systems—the consequences can range from trivial to catastrophic. It is essential that systems of accountability are established to address failures adequately. This involves not only identifying the reasons for failure but also having predefined protocols to manage negative outcomes effectively. For instance, in disaster response scenarios where predictive models might fail to accurately forecast a natural disaster, robust emergency measures must be pre-established and roles clearly defined to mitigate the impact effectively.

The realm of ethical considerations in predictive modeling is complex and multifaceted. Engaging with these considerations requires an interdisciplinary approach, incorporating insights from data science, ethics, law, and the specific fields affected by these forecasts. Furthermore, embracing a culture of responsible forecasting—

where ethical considerations are not afterthoughts but integral to predictive modeling processes—can help in harnessing the power of prediction in chaotic systems while safeguarding societal values and individual rights.

Chapter 8

Quantum Chaos: Bridging Micro and Macro Scales

This chapter explores the intersection of quantum mechanics and chaos theory, known as quantum chaos, which investigates the behavior of quantum systems that exhibit chaotic properties in their classical limits. It provides a foundational understanding of the subject by discussing key concepts of both fields and elaborating on their integration. The chapter also addresses the implications of quantum chaos for understanding fundamental processes at microscopic scales and its potential impact on the development of quantum technologies.

8.1 Introduction to Quantum Chaos

Quantum chaos, a term that sparks as much intrigue as it does confusion, is considered one of the most compelling subjects at the intersection of chaos theory and quantum mechanics. The journey into this fascinating field begins with an understanding that the classical concept of chaos—seemingly random and unpredictable behavior arising from deterministic systems that are highly sensitive to initial conditions—extends profoundly into the quantum domain. Yet, unlike classical chaos, which is defined by clear and well-documented paths toward unpredictable outcomes, quantum chaos concerns itself with the quantum behaviors in systems whose classical limits are known to be chaotic.

Transitioning from a classical to a quantum point of view, we must first establish that classical mechanics is wholly deterministic, and any chaotic behavior stems from the extreme sensitivity to initial conditions. This sensitivity leads to what is known in technical parlance as the "butterfly effect," where minute differences in initial conditions yield widely diverging outcomes, rendering long-term predictions impossible despite using precise, deterministic equations. Quantum mechanics, on the other hand, is fundamentally probabilistic. Herein lies the primary challenge: merging the deterministic chaos of classical systems with the intrinsic probabilistic nature of quantum systems. This merger is not straightforward, as the direct quantum analog of classical chaotic behavior does not exist due to the foundational principles of quantum theory such as the superposition principle and the Heisenberg uncertainty principle.

The concept of *quantum signatures of chaos* aims to bridge this divide. It refers to quantum characteristics in the statistics of energy levels and eigenfunctions of quantum systems whose classical analogs are chaotic. Research has shown distinct patterns in the level spacing of energy states in these systems—patterns that deviate markedly from those seen in systems classifiable as "quantum inte-

8.1. INTRODUCTION TO QUANTUM CHAOS

grable systems" (where there is no chaos). For instance, chaotic quantum systems typically exhibit level repulsion, where the spacing between consecutive energy levels follows a specific statistical distribution, contrasting sharply with the Poisson distribution observed in non-chaotic systems.

Analyzing these signatures is challenging yet rewarding. It necessitates a toolkit that blends both classical chaos insights and quantum mechanical analytics. One such tool is the semiclassical approach, where quantum effects are described using classical orbits, laying a direct link between quantum behavior and classical trajectories. Moreover, mathematical constructs like Gutzwiller's Trace Formula allow one to connect chaotic classical dynamics with quantum mechanical wave functions by extending over periodic orbits.

The study of quantum chaos not only expands our understanding of quantum mechanics but also brings a rigorous examination of chaos into areas previously unexplored by classical theories. As we delve through this chapter, focusing on the tools, techniques, impacts, and implications of quantum chaos, a clearer picture emerges of how chaos theory extends its reach into the quantum realm, influencing everything from atomic and molecular physics to burgeoning fields like quantum computing.

The ramifications of these studies are profound, influencing not only fundamental science but also practical applications in technology and computation. As we navigate through subsequent sections, keep in mind that each discussion builds upon these foundational insights, aiming to chart out the landscape where chaos meets the quantum world—a terrain rich with both complexity and considerable promise.

CHAPTER 8. QUANTUM CHAOS: BRIDGING MICRO AND MACRO SCALES

8.2 Foundational Concepts in Quantum Mechanics and Chaos Theory

Quantum mechanics and chaos theory, albeit strikingly different in their foundational premises, share the intrinsic nature of unpredictability in certain conditions. Here, we delve into the essential concepts of both fields that form the bedrock of understanding quantum chaos.

Quantum mechanics is anchored on the principle that particles at microscales (such as atoms and photons) exhibit dual characteristics of particles and waves. The wave function, denoted by $\Psi(x, t)$, is vital in quantum mechanics as it describes the probability amplitude of a particle's position and state in space and time. This function evolves according to the Schrödinger equation:

$$i\hbar \frac{\partial}{\partial t} \Psi(x, t) = \hat{H} \Psi(x, t)$$

where $\hat{H}$ is the Hamiltonian operator symbolizing the total energy of the system, and $i\hbar$ is the imaginary unit times the reduced Planck constant, embodying the quantum of action.

Another cornerstone of quantum mechanics is the concept of superposition, where a particle can exist simultaneously in multiple states until it is measured. This leads to the phenomenon known as quantum entanglement, where particles become interconnected and the state of one (no matter how distant) directly influences the state of another.

Chaos theory, conversely, emerges from the field of classical mechanics and dynamic systems where it describes

8.2. FOUNDATIONAL CONCEPTS IN QUANTUM MECHANICS AND CHAOS THEORY

systems sensitive to their initial conditions—a feature popularly known as the butterfly effect. Mathematically, chaos is typically associated with non-linear dynamical systems that can be described by deterministic equations yet exhibit unpredictable behaviors over time. An example is the well-known logistic map:

$$x_{n+1} = rx_n(1 - x_n)$$

This equation represents a simple model of population growth and demonstrates chaotic behavior for certain values of r.

In bridging these two theories, one inspects chaotic systems under quantum rules. While classical chaos discusses divergence of systems due to minute variations in initial conditions, quantum chaos explores how quantum effects can manifest in the macroscopic behavior of these systems. Since quantum mechanics fundamentally limits knowledge of initial states (Heisenberg's uncertainty principle), the transition from quantum to classical descriptions in chaotic systems entails a complex interplay of decoherence and entanglement.

An enlightening graphical representation to consider is the phase space plot for both classical and quantum systems. In classical mechanics, depicting trajectories in phase space commonly illustrates sensitivity to initial conditions. In quantum mechanics, however, such trajectories are forbidden due to Heisenberg's uncertainty principle. Instead, one might look at the evolution of probability densities or Wigner functions which encapsulate the quantum state akin to a phase space distribution in classic terms.

Furthermore, consider the role of eigenstates and energy levels, especially how high-energy levels often show a

CHAPTER 8. QUANTUM CHAOS: BRIDGING MICRO AND MACRO SCALES

semblance of chaotic behavior—termed 'quantum chaos'. Here's an example illustrated through a quantum billiard system, where a particle is confined within a boundary. The energy eigenstates of such a system can be visualized:

Energy Eigenstates in a Quantum Billiard System

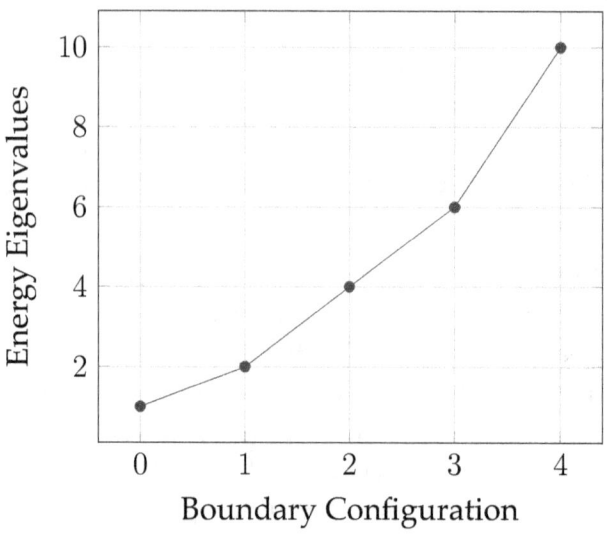

This plot serves as a rudimentary visualization illustrating how boundary conditions dramatically alter the energy states, a principle integral to understanding the effects on quantum chaos dynamics.

Through grasping these fundamental concepts of quantum mechanics and chaos theory, one prepares to delve deeper into how their interdependencies possibly explain various complex phenomena in quantum chaos. It's this intersection and synergy of theories that not only broadens our comprehension but also sparks novel insights into both realms.

8.3 Differences and Similarities with Classical Chaos

Diving deep into the domain of quantum chaos, it is critical to discern how it contrasts with and yet mirrors classical chaos. Classical chaos is known for sensitivity to initial conditions, often popularized as the "butterfly effect," where minuscule changes at the starting point can lead to vastly different outcomes. This phenomenon is described by the metric known as the Lyapunov exponent, which quantifies the rate of separation between infinitesimally close trajectories.

In quantum systems, however, the notion of trajectory is not well-defined due to the Heisenberg Uncertainty Principle, which states that the more precisely the position of some particle is determined, the less precisely its momentum can be known, and vice versa. This introduces a fundamental difference: quantum chaos does not involve sensitive dependence on initial conditions in the classical sense. Instead, quantum chaos studies the statistical properties of spectra, particularly how eigenvalues of quantum systems repel one another, a concept absent in classical systems.

Despite this fundamental difference, there are intriguing similarities. Both realms exhibit a transition from regular (predictable) to chaotic (unpredictable) behaviors as certain parameters are varied. For instance, in a classical double pendulum system, small changes in the length or mass of the pendulums can shift the system from a regular to a chaotic regime. Similarly, quantum systems like the hydrogen atom in a high magnetic field transition from regular to chaotic spectral statistics as the field strength increases.

CHAPTER 8. QUANTUM CHAOS: BRIDGING MICRO AND MACRO SCALES

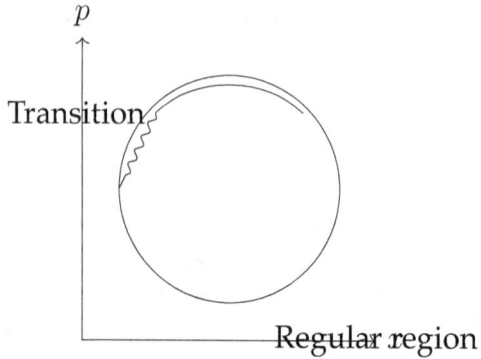

Furthermore, quantum chaos often relies on statistical measures derived from random matrix theory, used to predict the spacing of adjacent energy levels in quantum systems. These approaches echo tools used in classical chaos to investigate the spectrum of Lyapunov exponents across different dynamical regimes.

Another profound connection lies in Scarring of Wavefunctions—a concept in quantum mechanics wherein certain quantum states show a high probability density along classical unstable periodic orbits. This scarring effect serves as an evocative link between quantum and classical chaotic systems, suggesting that quantum systems remember their classical counterparts' chaotic features.

These discussions underscore that while the direct sensitivity to initial conditions may be muted in quantum chaos, the systems studied under this domain still share deep-rooted connections with classical chaos, particularly through their transition behavior and statistical properties of their respective spectra. These parallels and divergences not only enhance our understanding of chaos across different physical realms but also craft pathways for novel explorations in the overlap between quantum mechanics and classical dynamical systems.

8.4 Tools and Techniques for Analyzing Quantum Chaos

The analysis of quantum chaos hinges upon a sophisticated array of tools and techniques designed to probe and elucidate the counterintuitive behaviors witnessed at quantum scales. To fully grasp the nuances of chaos in quantum systems, one must employ both theoretical frameworks and practical methodologies that are capable of bridging the gap between quantum mechanics and chaotic dynamics.

At the forefront of these methodologies is the study of *quantum billiards*, an idealized model wherein particles move within a confined geometric boundary, interacting under the principles of quantum mechanics. This model is pivotal in understanding how chaotic trajectories evolve in systems that lack any apparent disordered potentials. To visualize these phenomenon, high-resolution spectral analysis is often employed, revealing how eigenvalues of the quantum system's Hamiltonian vary with changes in system parameters, a reflection of chaotic behavior.

Another crucial technique involves *Poincaré sections* and *phase space plots*, typically used in classical chaos but adapted for quantum systems. Modified for quantum use, these techniques involve studying the wavefunction's phase space distribution using tools like the Husimi or Wigner quasi-probability distributions. These distributions help highlight areas in phase space where the wavefunction exhibits complex, possibly chaotic behavior, allowing clearer distinction of quantum states' stability or instability.

Random Matrix Theory (RMT) offers another powerful framework for analyzing quantum chaos. Originally developed to address complex nuclei interactions, RMT has been widely adopted in quantum chaos to study statistical distributions of energy levels in quantum systems analogous to chaotic classical systems. When a quantum system's level spacing distribution follows that predicted by RMT, it often indicates underlying chaotic dynamics. Employing RMT helps in understanding the universal aspects of quantum systems exhibiting chaotic behavior independent of the specifics of their interactions.

Additionally, *time-series analysis* is applied through quantum state evolutions to identify signatures of chaos in time-dependent behaviors. Techniques like Fourier transforms are utilized to decompose the quantum state evolution into its frequency components. Examining these components provides insights into periodic orbits and their stability, which are crucial in discerning chaotic regions within the quantum dynamical space.

Technological advancements have also facilitated the direct simulation of quantum chaotic systems through computational models. Numerical simulations using platforms like MATLAB, Python's QuTiP (Quantum Toolbox in Python), or other specialized software enable detailed explorations into system dynamics under various initial conditions and perturbations. These simulations help validate theoretical predictions, offering a pragmatic look into systems too complex for analytical solutions alone.

Furthermore, *Quantum Lyapunov exponents* are increasingly researched as indicators of the rate at which quantum information spreads within a system, analogous to classical chaos indicators. Determining these exponents requires innovative computational techniques due to the inherently probabilistic nature of quantum mechanics versus deterministic classical mechanics.

8.5 Quantum Chaos in Atomic and Molecular Physics

Delving into the realm of atomic and molecular physics, quantum chaos emerges as a critical tool for unraveling the complexities inherent in the quantum behaviors of these systems. While traditional studies in this domain have heavily leaned towards linear, predictable models, recent advancements reveal that chaos theory can provide a more accurate depiction when dealing with highly excited states, often observed in atoms and molecules.

Understanding the implications of quantum chaos in this context involves examining how standard quantum mechanics, typically governed by smooth Hamiltonian evolution, intersects unpredictably when systems reach quantum limits similar to classical chaotic systems. Atoms, for instance, when driven by external fields at high intensities, display behaviors that deviate from what standard quantum mechanics would predict. This deviation is akin to the sensitivity to initial conditions—a hallmark of classical chaos.

Consider the hydrogen atom, a fundamental element in the panorama of quantum physics. When subjected to a strong electromagnetic field, the electron's trajectory within the atom begins to exhibit chaotic characteristics that cannot be described by regular perturbative methods. Instead, these conditions require non-perturbative techniques which consider the chaotic nature of the electron's motion at high energies.

CHAPTER 8. QUANTUM CHAOS: BRIDGING MICRO AND MACRO SCALES

A profound understanding can be garnered from the study of 'quantum scars' — an intriguing phenomenon observed in chaotic quantum systems where certain wavefunctions are found to concentrate along the paths of classical unstable periodic orbits. In molecular physics, similar observations have been made where molecules show increased reaction rates and unexpected energy transitions under conditions predicted by chaos theory. This is particularly evident in multi-electron dynamics where electron correlation plays a significant role and cannot be deciphered solely by independent particle approximations.

Probability Density of Quantum States in a Chaotic Atom

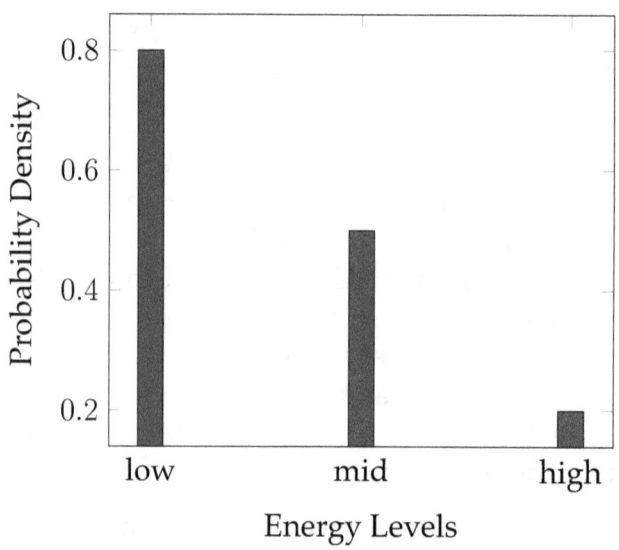

Quantum chaos also dramatically influences the stability and control of quantum states, a principle that directly impacts atomic and molecular computation and manipulation. Effective control over these states is foundational for developments in quantum computing and precision spectroscopy. Therefore, incorporating chaos theory into the design and implementation of these technologies can optimize their performance by accounting for potential chaotic disruptions.

From an educational perspective, introducing quantum chaos into the curriculum of atomic and molecular physics provides a more comprehensive understanding that breaks away from overly simplified models. The knowledge of how quantum systems can behave erratically and unpredictably enriches our understanding and enhances predictive capabilities in scientific research and applied physics.

8.6 Applications in Quantum Computing

Quantum computing represents one of the most thrilling frontiers in both computational science and quantum physics, with quantum chaos playing an increasingly recognized role in optimizing quantum algorithms and enhancing system stability. Here, we explore the confluence of quantum chaos and quantum computing, delving into the promising applications that arise from their intersection.

Quantum chaos provides a framework for understanding complex behaviors in quantum systems that can directly impact the development and performance of quantum computers. One of the primary benefits is the optimization of quantum gates and circuits. Quantum gates, akin to classical logic gates, are fundamental to quantum computing; they represent operations on quantum bits (qubits) that govern the computational processes. The chaotic nature of quantum dynamics can be harnessed to design more efficient quantum gates that are less prone to errors induced by environmental interactions and decoherence.

Consider a typical quantum circuit where coherence and fidelity are paramount. Chaotic systems' sensitivity to initial conditions can be utilized to amplify beneficial quantum effects while minimizing disruptive noise. For instance, applying principles of quantum chaos can lead the qubits to inhabit regions of the Hilbert space that are less susceptible to decoherence, a perennial challenge in practical quantum computing.

Another application worth exploring is in optimizing quantum search algorithms. Quantum algorithms, such as Grover's algorithm, already promise exponential speedups over their classical counterparts for searching through unsorted databases. Integrating quantum chaos into these algorithms helps distribute the amplitude of quantum states more evenly across the computing space. This uniformly distributed amplitude not only helps in faster convergence of the algorithm but also increases the stability and reliability of the outcomes.

Moreover, quantum chaos has proven instrumental in the development of random number generators using quantum systems. These generators are crucial for tasks ranging from cryptography to Monte Carlo simulations. By harnessing chaotic maps which have well-understood behavior in classical systems, researchers have developed analogous quantum chaotic systems. These systems exploit superposition and entanglement to produce sequences of random numbers that are fundamentally more unpredictable than their classical equivalents, thus enhancing security measures in quantum cryptography.

In addition to theoretical applications, experimental setups like the quantum kicked rotor (QKR) provide practical insights into the implementation of quantum chaos. The QKR has been used effectively to test fundamental concepts in quantum chaos like dynamical localization—a phenomenon akin to Anderson localization but occurring in momentum space. These studies not only further our understanding of chaotic phenomena in quantum mechanics but also pave the way for their practical applications in designing robust quantum computing architectures that are resistant to external perturbations.

Although the field is still burgeoning, the intersection of quantum chaos and computing holds potential for solving some of the most difficult problems in quantum mechanics, and information technology. By transcending classical limitations and harnessing chaotic behaviors at quantum scales, new paradigms in secure communications, efficient problem-solving algorithms, and robust error correction methods are likely to emerge.

8.7 Impact on the Understanding of Quantum Decoherence

Quantum decoherence represents a fundamental mechanism by which quantum systems lose their quantum characteristics and behave more classically. Within the scope of quantum chaos, the study of decoherence is poised uniquely—for it is here that we observe the transition of quantum information into classical noise which inadvertently characterizes the chaotic dynamics of quantum systems.

Quantum chaos extends the understanding of decoherence by demonstrating how chaotic dynamics accelerate the entanglement among the degrees of freedom within a quantum system, thereby hastening the decoherence process. This insight is pivotal because it links the stability of quantum information to the underlying chaotic or non-chaotic nature of the quantum system. A key example is found in quantum computing, where maintaining coherence is essential for the functionality of qubits. The emergence of chaotic behavior can lead to faster decoherence, thus posing significant challenges and guiding the design of more robust quantum systems.

Mathematically, the relationship between quantum chaos and decoherence can be understood through the correlation functions of chaotic systems, which decay exponentially—a hallmark of classical chaos mirrored in quantum systems. This decay can be approximated by:

$$C(t) = C(0)e^{-\lambda t}$$

where $C(t)$ is the correlation function at time t, $C(0)$ is the initial value, and λ is the rate of decay, indicative of the rate of entanglement and subsequently the rate of decoherence.

One illustrative model is the kicked rotor, a paradigmatic system used to study quantum chaos. When subjected to periodic delta kicks, the system transitions from regular to chaotic dynamics as the kick strength is varied. By integrating a measurement of coherence (fidelity) in each phase (regular or chaotic), we can quantitatively discern how chaotic dynamics affect decoherence rates. Here, fidelity measures the overlap between the initial state and the evolved state of the rotor, revealing faster decoherence under chaotic conditions.

Furthermore, studies utilizing random matrix theory have illuminated additional aspects concerning how eigenstate structures in quantum chaotic systems differ markedly from those in non-chaotic systems. These differences manifest notably in the density matrix's off-diagonal terms, which are significantly suppressed in chaotic systems, thereby indicating a faster approach to classicality.

To bring these concepts closer to practical relevance, consider the implications in developing technologies like quantum sensors and clocks, where precision is paramount. Understanding the impact of chaos on decoherence has led to optimized strategies for controlling environmental interactions—strategies vital for enhancing the performance and stability of these devices.

Although much progress has been made, the journey of exploring quantum chaos's influence on decoherence remains rich with opportunities. Forward-looking approaches are increasingly embracing sophisticated computational techniques to simulate quantum chaotic systems more effectively, anticipating their behavior to devise better countermeasures against unwanted decoherence.

To encapsulate, the incorporation of chaos theory into the realm of quantum mechanics through the lens of decoherence not only reframes our understanding but also enhances our ability to manipulate and utilize quantum systems in technology. Each step forward in this field brings us closer to mastering the delicate balance between harnessing quantum capabilities and mitigating chaos-induced vulnerabilities.

8.8 Future Research Directions in Quantum Chaos

Exploring the articulate dance between quantum mechanics and chaos theory presents a fertile ground for novel research, with several promising directions that can potentially reshape our understanding of both fundamental theories. The subtle harmonic convergence where quantum systems embody chaotic behaviors invitingly suggests a myriad of applications, from quantum computing to the fundamental nature of reality.

Enhancing Quantum Algorithms via Chaos Insights
One significant avenue for future research involves leveraging chaos theory to enhance the efficiency and robustness of quantum algorithms. While quantum computers offer exponential speed-ups for certain tasks, their practical implementation faces challenges such as error rates and decoherence. Introducing controlled chaotic dynamics into quantum algorithms might offer innovative ways to navigate these issues, potentially leading to breakthroughs in error correction schemes and computational speed. Studies focused on simulating chaotic systems using quantum computers have shown early promise, indicating that more in-depth exploration could yield transformative tools and techniques that capitalize on chaotic behaviors to optimize quantum computation.

Developing Quantum Chaos Experimental Platforms
Experimental investigations into quantum chaos are currently limited by available technologies. Future research should focus on the development and refinement of high-precision experimental setups capable of controlling and measuring quantum systems with sufficient detail to observe chaotic dynamics. This may involve enhancing

8.8. FUTURE RESEARCH DIRECTIONS IN QUANTUM CHAOS

technologies such as cold atom traps, superconducting qubits, and photonic circuits, which could be tailored to explore the rich structure of quantum chaos more comprehensively. By pushing the boundaries of experimental techniques, scientists can explore the subtle effects of quantum chaos in a controlled environment, leading to better theoretical models and applications.

Quantum Chaos in Biological Systems
Another intriguing research direction is the exploration of quantum chaos in biological systems. There is growing evidence suggesting that quantum effects might play a role in phenomena such as photosynthesis, avian navigation, and perhaps even human cognition. Understanding how chaotic dynamics influence these quantum processes could open up new areas of research at the intersection of quantum biology and chaos theory. This approach not only broadens our understanding of life at a molecular level but also may lead to novel bio-inspired technologies.

Implications for Fundamental Physics and Cosmology
On a more fundamental level, further investigation into the implications of quantum chaos for general theories of physics, particularly in quantum field theory and cosmology, remains an untapped area of significant potential. Quantum chaos could provide new insights into the behavior of high-energy systems, early universe conditions, or even black hole information paradoxes. Developing theoretical frameworks which incorporate elements of chaos theory into these domains might refine our understanding of universe's most fundamental aspects.

The connection between quantum systems at micro scales and chaotic dynamics provides a unique lens through which to view the universe. With targeted research efforts aligned toward uncovering these underpinnings, we can expect not only to advance our theoretical knowledge but also to discover practical applications that have yet to be envisioned. Each step forward in this domain will help to illuminate the enigmatic interplay between order and chaos, raising both challenges and opportunities for physicists and engineers alike.

Chapter 9

Chaos in Art and Culture: Influence and Inspiration

This chapter examines the profound influence of chaos theory on art and culture, illustrating how the concepts of unpredictability and complex patterns have inspired diverse artistic expressions across various media. It investigates the ways artists have harnessed chaotic principles to create visually compelling works and reflects on how these ideas resonate with broader cultural themes. By linking chaos theory with creative processes, the chapter highlights the valuable interplay between scientific insight and artistic exploration.

9.1 Exploring the Connection Between Chaos Theory and Art

Chaos theory, a compelling subset of mathematics and physical sciences, studies the behavior of dynamic systems highly sensitive to initial conditions—a presence denoted as the butterfly effect. This sensitivity dictates that small causes can have disproportionately major impacts, leading to seemingly random outcomes from deterministic systems. The parallels between this principle and the processes in artistic creation are not only remarkable but have also been significantly instrumental in shaping modern art. This connection offers a rich field of exploration where science and art converge, demonstrating that disciplines seemingly worlds apart can collaborate to enrich one another profoundly.

Artists, from painters to digital media specialists, often leverage chaotic dynamics to produce captivating, intricate patterns and visuals that may not have been achievable through traditional methods. The inherently unpredictable results that emerge from slight changes in a chaotic system reflect the essence of creativity—where minute variations in brushstroke, color blending, or pixel arrangement can lead to dramatically different outcomes.

Consider the work of artists like Jackson Pollock, known for his drip paintings which embody the randomness of chaos theory. Pollock's technique—allowing paint to drip and splatter on the canvas—creates an unpredictable array of shapes and colors. Each piece's exact outcome is highly sensitive to the environment's conditions, his movement, and the method of pouring paint, mirroring the unpredictable yet deterministic nature of chaotic systems.

In digital art, algorithmic and generative artists harness chaotic equations to create complex patterns and images that evolve from simple rules. These artists often employ fractal mathematics—a core concept in chaos theory—to generate elaborate designs that are infinitely complex and

9.1. EXPLORING THE CONNECTION BETWEEN CHAOS THEORY AND ART

often resemble forms found in nature, such as coastlines, mountains, or clouds. The Mandelbrot set, one of the most famous fractals, demonstrates how repeating a simple equation can yield infinitely complex designs, each iteration revealing greater intricacy.

Symbolic Representation of the Mandelbrot Set

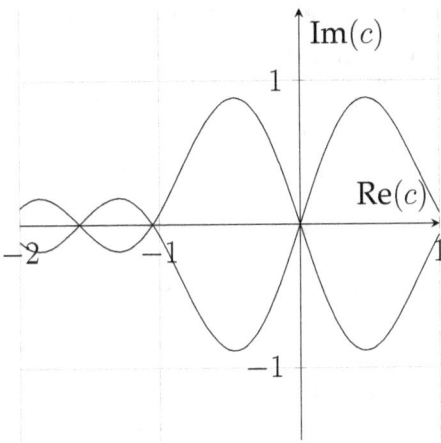

Figure 9.1: A symbolic and simplified graphical representation of the Mandelbrot set. The actual visualization requires numerical computation outside of LaTeX capabilities.

The creative process in art can mimic the sensitivity to initial conditions characteristic of chaotic systems. Artists might begin with a vision or a rough sketch – a set of initial conditions, in scientific parlance – and as they develop their artwork, small deviations can lead to significantly different aesthetic outcomes.

This overlap between chaos theory and artistic methodology is not only intriguing but also enlightening. It opens up new avenues for understanding the creativity inherent in both fields, suggesting that perhaps there exists a universal structure and order deeply embedded within what initially appears as chaos.

As this exploration unveils, the influence of chaos theory extends beyond scientific boundaries and mathematical equations—it permeates the soul of artistic expression. It challenges artists to experiment and innovate, using chaos not just as a concept but as a collaborator, to explore new territories in visual and digital media. By embracing the unpredictable nature of chaos, artists are able to push the boundaries of what is possible in art and open up new perspectives for viewers, ultimately expanding the scope of human creativity and understanding.

9.2 Visual Arts: Fractals and Patterns

The intricate dance between order and chaos in visual arts is vividly illustrated through the use of fractals and complex patterns. These elements, drawing directly from the mathematical roots of chaos theory, offer a fascinating lens through which artists explore the boundaries of beauty and complexity. This section delves into how fractals have not only influenced artistic methods but also enhanced the aesthetic appreciation of chaos in art.

Fractals, characterized by self-similarity across different scales, represent a core concept of chaos theory. Artists have utilized this concept to create multi-scale patterns that exhibit a mesmerizing combination of uniformity and variety. A prominent example can be the Mandelbrot set, discovered by Benoît Mandelbrot in 1980. It symbolizes an iconic image for chaos and has inspired a vast spectrum of artwork, from abstract paintings to digital art.

The application of fractals in visual arts is not limited to

9.2. VISUAL ARTS: FRACTALS AND PATTERNS

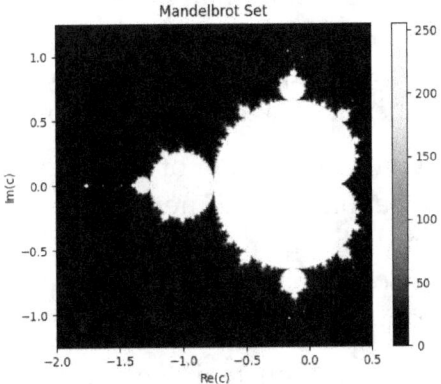

Artist	Fractal Dimension
Jackson Pollock	1.72

Table 9.1: Fractal dimensions of selected artists' works indicating the presence of chaotic elements in visual arts.

explicit renditions of mathematical sets. Artists like Jackson Pollock, whose drip paintings unwittingly echoed fractal patterns, demonstrate that the essence of chaos can percolate into art in unforeseen ways. Pollock's technique, involving the dripping and splattering of paint, resulted in intricate layers of lines and forms. Research conducted by physicist Richard Taylor in 1999 quantitatively analyzed these patterns, revealing a distinct fractal dimension consistent across many of Pollock's works. Such discoveries underscore the intrinsic, often unconscious connection between chaos theory and artistic expression.

Further exploring the application of chaos theory in visual arts, digital artists harness algorithms to generate complex patterns that emulate natural chaotic phenomena like turbulence, weather patterns, and vegetative growth. These computationally derived artworks not only mirror the unpredictability inherent in chaos but also invite viewers to explore the deeper connections between natural phenomena and artistic representation.

Additionally, the incorporation of fractal geometry in visual art extends beyond two-dimensional canvases to digital animation and film, providing a means to create more naturalistic landscapes and textures that appear infinitely detailed and realistically random.

As visual artists continue to integrate principles from chaos theory into their work, they challenge the traditional boundaries of artistic creation and redefine what it means to create art that reflects the natural world. Their efforts illuminate not just the beauty inherent in chaos but also the broader applicability of fractal and chaotic patterns across various forms of cultural expression. These artistic endeavors not only enhance our aesthetic experience but also deepen our understanding of the complexities within both nature and human perception.

9.3 Music and Chaos: Compositional Techniques

Delving into the realm of music, chaos theory provides a fascinating lens through which we can understand certain avant-garde compositional techniques. At its core, chaos in music involves the use of non-linear dynamics and unpredictable patterns that still maintain an underlying order. This juxtaposition of randomness and structure invites listeners into a complex auditory landscape, where expectations are both met and undermined.

9.3. MUSIC AND CHAOS: COMPOSITIONAL TECHNIQUES

One notable application of chaotic principles in music is through algorithmic composition. This method employs computer algorithms that incorporate elements of randomness and complexity, akin to chaotic systems observed in nature. A pioneering example can be seen in the works of Iannis Xenakis, who used stochastic processes—a form of mathematical randomness—to compose pieces that reflect the unpredictability yet structured reality of natural phenomena.

Composer	Technique
Iannis Xenakis	Stochastic music composition
John Cage	Indeterminate music using chance operations
Brian Eno	Generative music, systems that produce evolving patterns

John Cage, another influential figure, introduced the idea of indeterminacy in music. His compositions, such as the infamous "4'33"," where performers do not play their instruments for four minutes and thirty-three seconds, challenge the traditional notion of musical structure. Cage's use of I Ching in compositional decision-making injects an element of randomness that can be seen parallel to chaotic behavior where small changes in initial conditions can lead to vastly different outcomes.

Furthermore, the concept of fractals has also permeated musical composition. A fractal in music is a repeated pattern that varies in scale rather than in detail; hence, it may be manifest across different time scales or structural lev-

CHAPTER 9. CHAOS IN ART AND CULTURE: INFLUENCE AND INSPIRATION

els in a composition. Composers like György Ligeti utilize fractal geometry in their work to create intricate patterns that appear similar at various levels, resembling the recursive nature of fractals.

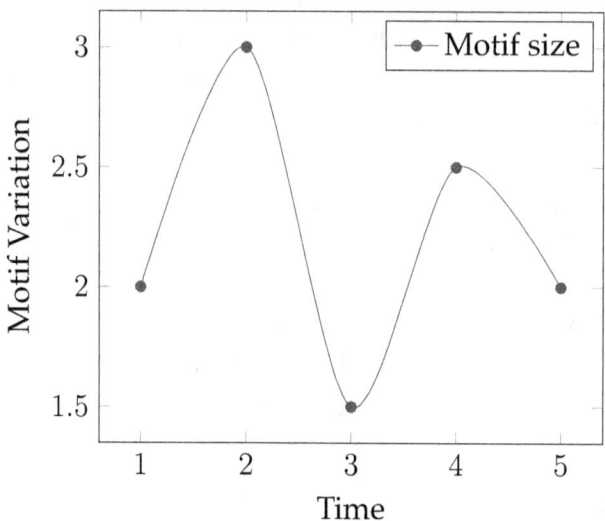

Fractal Music Visualization

Implementing chaos into musical systems doesn't end at composition alone; it extends to performance as well. The incorporation of real-time generative systems, as used by composers like Brian Eno in his ambient works, introduces live chaos. These systems generate music that is ever-different and responsive to initial conditions, reflecting the sensitive dependence on initial conditions found in chaos theory.

The bridge between chaos and music not only expands our understanding of composition and aesthetic enjoyment but also mirrors a larger cultural appreciation for complexity and subtlety in an era marked by technological advancements and increasing awareness of the complex systems that govern our world. Through these auditory explorations, composers provoke us to reconsider our perceptions of order, control, and the unpredictable nature of life itself.

9.4 Literature: Narratives and Themes of Chaos

Exploring the portrayal of chaos in literature unfolds a rich tapestry where authors encapsulate both the unpredictability inherent in human existence and the intricate order underlying apparent randomness. Chaotic systems, marked by their sensitivity to initial conditions—often illustrated by the butterfly effect—find their narrative parallel in literature, where minor events can cascade into significant, sometimes catastrophic outcomes.

Take, for instance, the works of Leo Tolstoy or Thomas Hardy, where characters' lives demonstrate profound shifts from seemingly trivial decisions or events. Hardy's "Tess of the d'Urbervilles" presents a stark exploration of chaos theory through Tess's life, where the cumulative effect of small, initial differences escalates into tragically large consequences, thus underlining the sensitivity to initial conditions.

Moreover, in modern literature, authors like Michael Crichton in "Jurassic Park" explicitly invoke chaos theory through their characters, such as mathematician Ian Malcolm, to critique the overreach of technological control and predict the unpredictability inherent in manipulating natural systems. Here, chaos is not only a thematic undercurrent but a vividly articulated principle that drives the narrative forward.

Narrative structure itself can mirror chaotic systems, demonstrated brilliantly in non-linear writings. Julio Cortázar's "Hopscotch" allows readers to approach the narrative in a non-linear fashion, jumping between chapters at will, effectively generating a unique experience with each reading, akin to observing a chaotic system from various initial conditions. This novel's structure reflects the unpredictability and diversity of outcome inherent in chaos theory.

Incorporating visuals derived from fractals in literature also illuminates connections with chaos. For example, images accompanying a text or shaped poetry often utilize fractal-like patterns to enhance and complement the chaotic themes of the writing. Consider Mark Z. Danielewski's "House of Leaves", where the chaotic layout of text across the pages intensifies the reader's experience of disorientation and unpredictability.

Furthermore, chaotic dynamics are not just thematic elements but can drive plot development. In "The Sound and the Fury" by William Faulkner, the fragmented and nonlinear narrative reflects the psychological turmoil and decay within the Compson family. Through shifting perspectives and timelines, Faulkner encapsulates the essence of chaos, both in form and content—every reading potentially offering a new interpretation or insight.

As readers engage with such texts, they witness how minor narrative shifts or stylistic choices can have profound effects on their understanding and interpretation, much like how slight changes in initial conditions can vastly alter the outcome in a chaotic system. This engagement not only enriches their reading experience but deepens their appreciation of the complex interplay between order and chaos.

This exploration through literature not only highlights the creative utilization of chaos theory but also underscores a universal resonance with the chaotic nature of human life—where order and predictability are continuously negotiated against the backdrop of chaos. Through these narratives and themes, literature provides a profound reflection on and engagement with chaos, offering readers a lens to view both the story and their realities anew.

9.5 Architecture: Incorporating Nonlinear Dynamics

The incorporation of nonlinear dynamics into architecture represents a groundbreaking paradigm where the predictability of classical designs gives way to innovative, adaptive structures inspired by the principles of chaos theory. This approach has yielded not only aesthetically unique designs but also sparked debates on the relationship between environment, usability, and architectural form.

First and foremost, nonlinear dynamics in architecture often manifest through the use of organic, seemingly random patterns that mimic the irregular, yet patterned, complexity found in nature. These patterns are not merely decorative but integral to the building's functionality and interaction with environmental factors. For example, the design of the Eden Project in Cornwall, England, by Nicholas Grimshaw, is an application of such principles. The project features a series of interconnected domes that employ a hex-tri-hex space frame pattern mimicking bubble geometry. This not only maximizes structural efficiency but also harmonizes with the natural environment.

To illustrate this further, consider how these principles apply to wall configurations and surface patterns. Architects might use algorithm-based tools to generate wall surfaces that respond adaptively to sunlight, enhancing natural lighting inside the building while minimizing solar heat gain. The use of parametric design software allows for these complex calculations and designs that respond in real-time to changing environmental conditions, effectively incorporating elements of chaos into a harmonious design.

Moreover, this section delves into the integration of fractal geometry into architectural elements. Fractals are structures that exhibit self-similarity at different scales, a concept derived directly from chaos theory. By incorporating fractal elements into facades or structural components, architects achieve designs that are not only visually captivating but also structurally sound. A prominent example is the Federation Square in Melbourne, Australia, designed by Lab Architecture Studio in collaboration with Bates Smart, which features an intricate fractal façade that creates varying shadow patterns, altering the building's appearance and interaction with sunlight throughout the day.

Another significant aspect of incorporating nonlinear dynamics is the emphasis on designing structures that can adapt to unpredictable changes, such as shifting demographics or climatic conditions. Adaptive architecture uses materials and technologies that enable buildings to respond dynamically to their occupants and environments. For instance, kinetic facades made up of modules that can shift based on external weather patterns and internal usage create a dynamic interface that changes according to functional needs and energy optimization strategies.

It is noteworthy that while the aesthetic and functional aspects are prominent, the ethical and philosophical implications also play a crucial role. The adoption of chaos-centric designs challenges traditional perceptions of architectural form and function, proposing a model where buildings are seen not as static entities but as embodiments of change and adaptation, reflecting the fluid nature of human existence and our interactive relationship with our surroundings.

As architects continue to explore and refine the integration of nonlinear dynamics in building design, they not only revolutionize how structures are conceptualized and realized but also enhance our understanding of architecture as an art form that mirrors the complex patterns of life itself, embracing chaos not as a challenge to overcome, but as an inspiration to evolve.

9.6 Cinema: Depicting Chaos and Complexity

Cinema, as a dynamic form of visual storytelling, offers a unique platform for portraying the nuanced facets of chaos and complexity. Films utilize a blend of narrative, editing, and visual effects to simulate real-life scenarios influenced by chaotic principles, engaging audiences in a visual exploration of instability and unpredictability.

One of the primary cinematic techniques used to depict chaos is nonlinear storytelling. Unlike traditional narratives that follow a straightforward, causal sequence, nonlinear films often present events out of order, mimic the unpredictable behavior observed in chaotic systems, and challenge the viewer's perception of time and reality. This narrative style can be seen in films such as 'Pulp Fiction' by Quentin Tarantino and 'Memento' by Christopher Nolan, where the disjointed sequence fosters a compelling depiction of chaos.

Moreover, the use of fractal imagery in film visuals also demonstrates the marked influence of chaos theory. Fractals, with their self-similar pattern repeated at every scale, have been effective in scenes that aim to portray complexity in a visually arresting manner. The film 'The Tree of Life' by Terrence Malick, for example, incorporates fractal visuals to represent the intricate and interconnected nature of life, reflecting chaos theory's premise that simple rules can govern complex systems.

Cinematographers and directors often employ rapid editing techniques to create a sense of disarray and tension that mirrors chaotic environments. Quick cuts and abrupt shifts between scenes or time frames replicate the jarring and unpredictable nature of chaotic systems. This technique is effectively used in action and thriller genres to enhance the chaotic atmosphere that drives the adrenaline-fueled narratives.

Sound design in films also plays a crucial role in conveying chaos. Composers may use atonal music or complex, layered soundscapes that defy traditional harmonic patterns, which can evoke feelings of unease or disorientation in the audience. This aural representation of chaos complements the visual and narrative layers, creating a cohesive chaotic environment that immerses the viewer.

In addition to these techniques, some films explore chaos

theory more explicitly through their themes and dialogues. Films such as 'Jurassic Park' discuss chaos theory directly through the character Ian Malcolm, a mathematician specializing in chaos theory, who explains how unpredictable events dominate the behavior of the park's dinosaur population. This not only serves as a plot mechanism but also subtly educates the audience about the principles of chaos theory.

The cinematic portrayal of chaos and complexity not only entertains but also enlightens, pushing the boundaries of traditional filmmaking to explore and reflect on the unpredictability inherent in our world. Through innovative narrative structures, visuals, editing, and sound design, filmmakers continue to capture the essence of chaos theory, making it accessible and engaging for a broader audience. These films encourage viewers to contemplate the role of unpredictability in life and appreciate the beauty that can emerge from chaotic systems.

9.7 Cultural Impacts: Perceptions of Chaos in Society

The intricate relationship between chaos theory and societal perceptions offers a fascinating vista into how scientific ideas permeate popular consciousness and influence cultural constructs. This section delves into the multifaceted ways in which the concepts derived from chaos theory have echoed through various aspects of society, revealing a compelling narrative about how humans interpret the notion of chaos.

Chaos theory, originally a domain confined to the scientific community, has transcended its initial boundaries to spark discussions in realms ranging from economics and sociology to philosophy and religious studies. It challenges the traditional belief in a deterministic universe, instead proposing a system where small variations in initial conditions can lead to vastly different outcomes. This idea has profound implications on our understanding of free will, fate, and the very fabric of reality, thus captivating the cultural imagination.

One of the most prominent manifestations of chaos in society is visible in economic modeling. Traditional economic models often struggled to predict crises or explain sudden market changes until chaos theory provided a framework for understanding complex systems. The notion that small triggers can lead to significant economic events has reshaped how economists view market dynamics and instability. This understanding has filtered down into public consciousness, subtly altering how ordinary people perceive economic fluctuations and their own roles within these systems.

In sociology, chaos theory has been employed to explain social dynamics and structures. It illustrates how small-scale interactions can lead to emergent behavior on a larger scale, a concept that has been used to describe phenomena ranging from crowd behavior to the spread of global cultures. For instance, the rapid, unpredictable spread of information on social media platforms exemplifies chaotic dynamics, with minor incidents sometimes sparking significant social movements.

Philosophically and ethically, the implications of chaos theory provoke debates about determinism and randomness. These discussions resonate deeply within cultural discourses, particularly in the context of existential questions about human existence and the unpredictability

9.7. CULTURAL IMPACTS: PERCEPTIONS OF CHAOS IN SOCIETY

of life paths. The random, yet deterministic nature of chaotic systems challenges conventional notions of morality and justice, pushing philosophers and ethicists to revisit age-old debates under this new light.

Visually, we see chaos theory reflected in media and advertising, where the appeal to complex and dynamic designs mimic chaotic attractors and patterns, capturing viewers' attention and provoking engagement through unpredictability. This aesthetic of complexity mirrors a societal shift towards valuing dynamic, non-linear thinking over static, linear approaches.

Religious and spiritual discourses also engage with chaos theory. Some interpret the unpredictable yet deterministic elements of chaotic systems as a modern expression of ancient beliefs in interconnectedness and divine intervention. This blend of science and spirituality offers a unique lens through which different cultures interpret their experiences and existence.

Let's consider a chart revealing the interest over time in chaos theory concepts across various societal sectors. The x-axis could represent time in years while the y-axis could denote increasing societal interest measured through metrics such as academic publications, media mentions, art exhibitions, etc.

CHAPTER 9. CHAOS IN ART AND CULTURE: INFLUENCE AND INSPIRATION

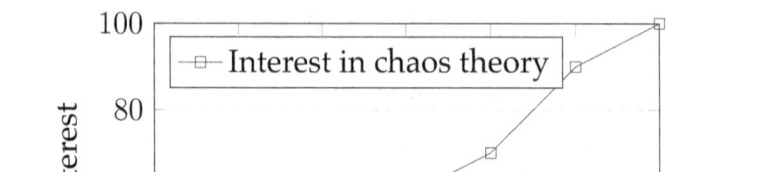

In this dynamic exploration of chaos theory within cultural frameworks, we observe not just the evolution of a scientific concept but its profound ability to influence and mirror the complexities of human society. Through its lens, we encounter a nuanced portrayal of human endeavor and creativity in navigating the unpredictable torrents of life's chaos.

9.8 Artistic Expression as a Reflection of Chaotic Concepts

Artists across various disciplines have consistently drawn from scientific principles to inform their work, but perhaps none so intriguingly as those inspired by chaos theory. This section delves into the myriad ways that chaotic concepts manifest within different forms of artistic expression, illustrating the deep and often unex-

9.8. ARTISTIC EXPRESSION AS A REFLECTION OF CHAOTIC CONCEPTS

pected connections between chaos theory and the arts.

Chaos theory, essentially, deals with the unpredictability found within dynamic systems. In art, this unpredictability can be mirrored in the spontaneity of brush strokes, the unstructured forms of jazz improvisation, or the complex narrative layers found in contemporary literature. Each of these examples demonstrates an embrace of the inherent unpredictability and complex patterns, which are hallmark traits of chaos.

In **visual arts**, painters like Jackson Pollock have famously utilized techniques that resonate deeply with chaotic dynamics. Pollock's drip paintings can be seen as a physical manifestation of chaotic motion, where the highly unpredictable movement of the paint is controlled by the artist's subtle movements and choices—yet the outcome remains fundamentally unforeseen. These paintings can be visualized as an artist's controlled experiment in chaos, where traditional tools for creating predictable patterns are eschewed in favor of methods that highlight complexity and randomness.

Spectral Analysis of a Pollock's Drip Painting

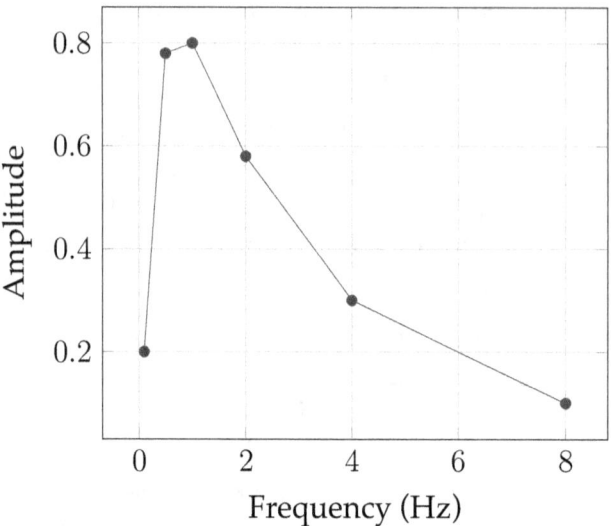

In **modern literature**, authors such as Julio Cortázar and Thomas Pynchon have woven chaos theory into their narrative structures, utilizing fragmented, nonlinear storylines that defy traditional narrative expectations and mimic chaotic systems by presenting readers with events that appear random but are interconnected beneath the surface. This incorporation into narrative structure reflects a deeper understanding of the universe's chaotic nature and challenges readers to find underlying order.

Music also mirrors chaos theory, particularly through genres like jazz and electronic music. Composers and performers in these styles often employ improvisation—a technique that appears surface-level chaotic but is deeply influenced by underlying rules of harmony, rhythm, and melody expression. Miles Davis's "Bitches Brew" is an exemplary record that uses chaotic musical structures to explore new depths of musical innovation.

9.8. ARTISTIC EXPRESSION AS A REFLECTION OF CHAOTIC CONCEPTS

```
Tuning Map for "Bitches Brew" Improvisation:
- Scale: Mixolydian with frequent chromatic shifts
- Rhythm: Varying time signatures 4/4, 7/8, 13/16
- Harmony: Non-traditional harmonics play with dissonance
```

Moreover, in the sphere of **digital art**, algorithmic and generative art forms use software code to create artwork that is inherently unpredictable. These digital artists program a set of rules or algorithms, which then generate art—similar to establishing initial conditions and parameters in a chaotic system and observing the unpredictable outcomes.

Finally, integrating chaos theory into artistic expression does not solely reflect a fascination with unpredictability; rather, it also symbolizes an acknowledgment of the complexity inherent in human experience. As we explore these expressions of chaos in art, we come closer to grappling with the fundamental nature of existence itself, full of unpredictable events and patterns that seem random but hold deeper interconnections.

Artistic engagement with chaos thus acts as both a mirror and a probe: reflecting our world's complex unpredictability while simultaneously exploring its mysterious depths.

CHAPTER 9. CHAOS IN ART AND CULTURE: INFLUENCE AND INSPIRATION

Chapter 10

The Future of Chaos: Emerging Trends and Theories

This chapter focuses on the evolving landscape of chaos theory, identifying emerging trends and novel theoretical developments that promise to expand the understanding and applications of chaos. It discusses how advancements in computational power, interdisciplinary research, and innovative analytical methods are driving new discoveries and insights. The chapter also speculates on the future implications of these developments, both for scientific research and practical applications, highlighting the dynamic and ever-changing nature of chaos theory.

CHAPTER 10. THE FUTURE OF CHAOS: EMERGING TRENDS AND THEORIES

10.1 Current State of Chaos Theory Research

One cannot delve into the future of chaos theory without a comprehensive understanding of its current state. The past few decades have seen profound advancements in chaos theory, particularly in its analytical capabilities which have been enhanced by the exponential growth of computational resources.

Comprehensive Computational Models: Recent research has leveraged high-powered computational technology not just to simulate but also to predict chaotic systems with unprecedented precision. Utilizing algorithms derived from machine learning and data science, researchers are now Modeling dynamic systems in environments as varied as weather systems, stock markets, and neural activities in the brain. For example, the use of recurrent neural networks (RNNs) has provided insights into predicting chaotic system evolutions which were once thought unpredictable.

Interdisciplinary Approach: Cross-disciplinary fusion forms one of the pillars of current chaos theory research. By merging concepts from mathematics, physics, biology, and computer science, a much richer understanding of chaotic behaviors across different systems has been realized. For instance, concepts originally used in weather prediction models have found their way into understanding the financial markets' irregularities and vice versa.

Advancements in Nonlinear Dynamics: The exploration into nonlinear systems, a core component of chaos theory, has seen significant breakthroughs. Recent studies employ nonlinear dynamics to uncover the complex be-

10.1. CURRENT STATE OF CHAOS THEORY RESEARCH

havior of various physical and biological systems. The non-linear analysis of electroencephalogram (EEG) data, for example, helps detect chaotic patterns that may indicate neurological abnormalities like seizures.

By harnessing the power of modern computational methods and interdisciplinary approaches, the domain of chaos theory has moved beyond mere theoretical curiosity. Researchers are increasingly able to not only describe chaotic systems but to also control them. Techniques such as control theory, applied in conjunction with chaos theory, have enabled scientists to steer chaotic systems towards desired states, hence mitigating potentially destructive behaviors inherent in uncontrolled chaotic systems.

Furthermore, continuous innovations in high-performance computing platforms provide a fertile ground for the next wave of discoveries. With supercomputers and quantum computing, the simulation and modeling capabilities at our disposal continue to grow exponentially. Such tools not only enhance our understanding but also push the limits of what chaos theory can achieve, transitioning from predictive modeling to prescriptive usage.

The state of chaos theory today is vibrant and rapidly evolving, underpinned by a strong computational backbone and enriched by cross-disciplinary strategies. It stands on the brink of even greater discoveries and applications, poised to reveal deeper insights into the nature of chaos in diverse and complex systems.

Now, as we pivot towards the intricate interconnections within chaotic systems, the scope of inquiry broadens and the depth of understanding deepens, setting the stage for the paradigms discussed in the subsequent sections. The symbiosis between technological advancements and theoretical models is reshaping our approach towards comprehending and manipulating chaos, heralding a new era in scientific exploration where the once-imperceptible patterns in chaos are not only observed but also harnessed.

10.2 Emerging Technologies and Their Impact on Chaos Theory

The evolution of technology has perennially acted as a catalyst in the development and reframing of scientific theories. In the case of chaos theory, the rise of emerging technologies such as Quantum Computing, Artificial Intelligence (AI), and enhanced computational models significantly impacts how researchers understand chaotic systems and predict their behavior.

Quantum Computing: This emergent technology is revolutionizing multiple areas of science and engineering by offering computational powers exponentially greater than classical computers. Quantum computing leverages quantum bits to perform complex calculations at high speeds, enabling scientists to simulate chaotic systems more accurately. For instance, the quantum properties allow researchers to model highly complex dynamical systems that were previously out of reach due to computational limitations. This advancement provides deeper insights into previously obscure aspects of chaos theory, such as the minutiae of quantum fluctuations within chaotic systems.

Artificial Intelligence and Machine Learning: The application of AI and machine learning in chaos theory opens up unprecedented avenues for predicting chaotic behavior in systems where traditional models fail. Through the use of neural networks and deep learning algorithms, AI can recognize patterns within chaotic systems, making it possible to predict changes and potentially control these systems with a higher degree of precision. For example, researchers have developed AI models that can predict the chaotic behavior of weather systems, which are notoriously difficult to forecast due to their sensitive dependence on initial conditions. Moreover, AI's ability to sift through vast amounts of data rapidly allows for a more nuanced understanding of the underlying structure in chaotic systems.

New Computational Models: Enhanced computational techniques such as cloud computing and high-performance computing (HPC) are pivotal in handling the vast computational needs required for analyzing chaotic systems. These platforms provide the necessary infrastructure to perform extensive simulations and process large datasets essential for chaos theory research. For instance, HPC enables the simulation of complex fluid dynamics problems, which are central to understanding turbulence, a classical example of a chaotic system.

To visualize how quantum computing influences the analysis of chaotic systems, consider the following diagram implemented in *TikZ*, illustrating a hypothetical quantum circuit used for simulating a chaotic system:

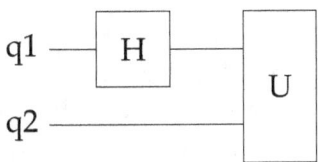

This simple representation depicts a two-qubit system where 'H' denotes a Hadamard gate applying a superposition state, and 'U' represents a unitary operation that could encapsulate a chaotic evolution under certain conditions.

Emerging technologies are not merely tools but transformative agents that redefine the boundaries and capabilities of chaos theory, pushing it towards more accurate predictions and broader applications across different fields. These technologies foster novel insights and methodologies that continue to enrich the understanding of chaos in increasingly profound ways.

10.3 Integration of Chaos Theory Across Disciplines

The expansive reach of chaos theory into various fields of study underscores its universality and adaptability. Acknowledging this, let's navigate the confluence where chaos theory integrates across multiple disciplines, detailing specific instances and elucidating how it enriches each field.

In meteorology, chaos theory has been a cornerstone for understanding weather patterns and atmospheric phenomena. The theory's inception began when Edward Lorenz discovered that small differences in a dynamic system such as the atmosphere could lead to vast differences in outcome, founded on his initial work on weather prediction models in the 1960s. Today, modern meteorologists utilize chaos theory to enhance their prediction models, accounting for the inherent unpredictability in weather systems. This adaptation has led to a deeper understanding of weather dynamics and improved the accuracy of short-term weather forecasts.

10.3. INTEGRATION OF CHAOS THEORY ACROSS DISCIPLINES

Transitioning to ecology, chaos theory aids in comprehending population dynamics in ecosystems. It helps ecologists analyze complex patterns such as population fluctuations and extinction events that may appear random but are in fact deterministic, governed by chaotic dynamics. This has crucial implications for conservation biology, enabling scientists to refine strategies for managing wildlife reserves, preventing species extinction, and understanding the chaotic nature of ecological responses to human-induced changes.

In the field of economics, the application of chaos theory is equally transformative. Economists apply chaotic models to analyze market dynamics and predict economic cycles. The non-linear models help in understanding the volatile behaviors seen in stock markets and commodities trading. By considering the chaotic elements of human behavior and external economic shocks, economists can forge better models for risk management and economic forecasting.

Health sciences, particularly neuroscience and cardiology, also benefit from chaos theory. In neuroscience, the theory explores the neuron's firing dynamics, offering insights into brain functionality and disorders such as epilepsy, where the seemingly unpredictable seizures are analyzed through chaotic models. In cardiology, chaos theory helps in understanding heart arrhythmias by examining the non-linear dynamics of heart rhythms. Such insights are pivotal for developing treatment strategies that mimic natural heart rhythms rather than suppressing them.

Moving into engineering, chaos theory finds relevance in system control and signal processing. Engineers incorporate chaos to improve the robustness and efficiency of

CHAPTER 10. THE FUTURE OF CHAOS: EMERGING TRENDS AND THEORIES

systems. For instance, chaotic maps are used to design random number generators essential in secure communications. Furthermore, the application of chaos in mechanical engineering assists in predicting system failures and enhancing the stability of structures subject to dynamic stresses.

Visualizing the integration along different disciplines is also facilitated through tools like dynamic systems mapping and non-linear differential equations applied ubiquitously across disciplines. The following example provides a simple depiction of a non-linear system modeled by a differential equation:

$$\frac{dx}{dt} = \sigma(y - x),$$

$$\frac{dy}{dt} = x(\rho - z) - y,$$

$$\frac{dz}{dt} = xy - \beta z.$$

These equations, resembling the famous Lorenz equations, highlight the convenient representation of complex chaotic systems found in multiple disciplines from weather forecasting to economic models.

As chaos theory continues to penetrate these diverse fields, its role becomes crucial not only in theoretical advancements but also in practical applications. It fosters new methodologies and theories that contribute afresh to each scientific enclave it touches, showing that even in diversity, there is a singular underlying order shaped by chaos. The interlinkage of chaos across various scientific terrains not only enhances our understanding but also exemplifies the unity in science—a myriad of disciplines connected through the thread of chaos.

10.4 New Mathematical Tools and Approaches

The exploration of chaos theory ushers us into an era where traditional mathematical methods often fall short of providing the clarity required to discern underlying patterns in chaotic systems. It is, therefore, imperative to develop new mathematical tools and approaches that enhance our ability to model, analyze, and predict chaotic phenomena. The following discourse unfolds some of the novel mathematical strategies that are enriching our understanding of chaos.

Topological Data Analysis (TDA) has emerged as a powerful tool in understanding the shape of data. In chaos theory, the topology of trajectories in phase space—places where the system's states evolve—can be intricate. TDA helps in capturing these complex structures in a manner that is invariant under continuous deformations. The utility of persistent homology, a mainstay technique of TDA, lies in its ability to extract features from the data at multiple spatial scales, thus identifying cycles and holes that persist across various parameter settings. This approach is particularly useful for distinguishing noise from true chaotic behavior in experimental data sets.

Network Theory provides a framework to study the interconnectivity within a chaotic system. By modeling chaotic systems as complex networks, scientists can examine how interactions between different components lead to emergent behavior. For instance, nodes in the network could represent different states of the dynamical system and edges could signify transitions between these states driven by the system's dynamics. Network measures such as node centrality or community structure can then illuminate aspects of the underlying chaos. For example, regions with high centrality might identify crucial states that could influence the system's trajectory towards chaos.

Machine Learning and Deep Learning Techniques are also profoundly transforming how we approach chaotic systems. These methods excel in identifying patterns and can be trained to forecast chaotic systems' future states with remarkable accuracy. Deep learning architectures, like Long Short-Term Memory (LSTM) networks, have been adept at modeling time series data, capturing the temporal dynamics integral to chaotic systems. Moreover, neural network-based frameworks have been developed to learn the governing equations of chaotic systems directly from data, a technique often referred to as *data-driven discovery* of dynamical systems.

Symbolic Dynamics is another fascinating area that reduces the dynamics of a chaotic system to sequences of symbols according to how trajectories traverse partitions of the phase space. This discretization converts a potentially incomprehensible chaotic trajectory into a comprehensible sequence of symbols, making it easier to analyze and predict system behaviors. It bridges a connection between qualitative dynamical descriptions and quantitative state transitions, providing a scaffold for understanding complexity in chaos.

Renewal of Interest in Nonlinear Differential Equations, particularly in non-standard forms, is enabling deeper explorations into chaos. These equations often defy analytical solutions, but with today's computational power and numerical methods like pseudo-spectral techniques, their behaviors are being charted with unprecedented detail. Such investigations not only offer insights into specific systems but also guide the general theory underlying chaotic dynamics.

As these tools develop and interlace, they drive forward our comprehension of chaos, guiding us through the previously uncharted waters of complex systems. Their deployment not only expands the horizon of chaos theory but also empowers a broader transdisciplinary approach to science, harnessing chaos for advancement in various fields. These tools do not stand alone; they intertwine and interact, each enhancing the impact of the other and leading to a richer, more precise understanding of chaotic systems.

It's clear that as we press forward, these innovative approaches will continue not only to reshape our understanding of chaos but also redefine the very methodologies by which we explore complex systems. The dialogue between researchers in mathematics, physics, computer science, and beyond hints at even greater synergies and innovations—promising a thrilling chapter ahead for chaos theory.

10.5 Future Applications in Science and Engineering

The potential applications of chaos theory in science and engineering are broad and impactfully diverse, with emerging opportunities endowing researchers and practitioners with new tools to solve complex, dynamic problems. The advancement of this theory into practical realms marks a bridge between abstract mathematical concepts and tangible innovations that could redefine our technological landscape.

CHAPTER 10. THE FUTURE OF CHAOS: EMERGING TRENDS AND THEORIES

Firstly, the use of chaos-based algorithms in the realm of artificial intelligence and machine learning is particularly promising. As computational models grow more sophisticated, integrating chaos theory can enhance the ability to model and predict nonlinear and dynamic systems. For instance, neural networks incorporating chaotic dynamics can generate more robust models for predicting weather patterns, financial markets, or complex biological processes. These systems benefit from chaos theory's sensitivity to initial conditions, which can be exploited to improve the accuracy and efficiency of predictive algorithms in environments where data is ephemeral or noisy.

In the field of robotics, chaotic control strategies can be utilized to create more adaptable and resilient robots. Traditional robots operate effectively in structured environments, where variables are controlled and predictable. However, by integrating chaos theory into robotic control systems, robots can achieve improved adaptability in unpredictable or complex environments. This capability is particularly critical in disaster response scenarios or operations in irregular terrains, where pre-programmed behaviors are insufficient.

Furthermore, chaos theory finds its application in the optimization of engineering systems such as power grids and transportation networks. Traditional methods often fall short when dealing with complex, fluctuating demand patterns, but chaotic modeling can provide solutions that adjust more dynamically to changing conditions. For example, applying chaotic sequences to model traffic flow can lead to the development of adaptive traffic control systems that optimize flow and reduce congestion based on real-time conditions.

Energy systems, specifically renewable energy integra-

10.5. FUTURE APPLICATIONS IN SCIENCE AND ENGINEERING

tion, also stand to gain from chaos theory applications. The inherently unpredictable nature of sources like wind and solar power can be better managed using chaotic models, which can simulate and predict energy production variability. This enhances grid stability and efficiency, allowing for a higher proportion of renewable energy in the power mix.

In the medical field, chaos theory is revolutionizing how we understand and approach complex systems such as the human cardiovascular system or brain function. Researchers are applying chaotic models to study the irregularities in heart rhythms to predict cardiac diseases or to model neural activities for insights into disorders like epilepsy or Alzheimer's. The detailed understanding of these chaotic patterns can lead to better diagnostic tools and treatment plans tailored to individual patients' unique system dynamics.

While moving forward, integration of chaos theory into these disciplines will necessitate not only advanced mathematical skills but also collaborative approaches that merge insights from various scientific and engineering fields. The capability to master and manipulate chaotic systems opens up a new frontier in designing solutions that are as dynamic and unpredictable as the problems they aim to solve. As we continue to uncover the layers of complexity in chaotic systems, the future holds a promise of unprecedented applications that could fundamentally alter our interaction with technology and nature.

With these burgeoning applications on the horizon, the convergence of chaos theory with science and engineering is not just a possibility but an inevitable shift that will carry significant implications for innovation. The realm where chaos reigns might just be the fertile ground on

CHAPTER 10. THE FUTURE OF CHAOS: EMERGING TRENDS AND THEORIES

which the future of technological advancements and scientific breakthroughs will thrive.

10.6 Evolving Theories in Quantum Chaos

Quantum chaos, a term often used to describe the complex behavior of quantum systems that exhibit classical chaos in the classical limit, is a vibrant frontier in theoretical physics. The exploration to understand how classical chaos theory translates into the quantum regime has unveiled a plethora of research questions and methodologies. Quantum systems, unlike their classical counterparts, do not conform to the predictable trajectories of classical mechanics; instead, they are governed by probability amplitudes, leading to a non-intuitive blend of stochastic and deterministic theories.

The seminal work on the quantum chaos theory began with the study of spectral statistics of quantum systems whose classical analogs exhibit chaotic behavior. One of the pivotal findings in early quantum chaos theory was the discovery of the level-spacing distribution of eigenvalues of quantum systems, which follows the Wigner-Dyson statistics rather than Poisson statistics, indicating a profound difference in organization and structure as opposed to regular quantum systems.

Recent advancements in quantum chaos are fueled by innovative computational techniques and experimental designs. A prime example is the development of the kicked rotor model, a quantum analogy to the classical kicked rotor that exhibits chaotic behavior. The kicked quantum rotor demonstrates sensitivity to initial conditions—a hall-

10.6. EVOLVING THEORIES IN QUANTUM CHAOS

mark of chaotic systems—in a way that aligns quantum mechanical effects with classical chaos through phenomena like dynamical localization, a quantum effect that has no classical counterpart but parallels Anderson localization in disordered systems.

Advancements in quantum computing have also opened new pathways for exploring quantum chaos. Quantum simulators, capable of manipulating small-scale quantum systems, enable researchers to create and study models that were previously impractical to explore. This has led to groundbreaking investigations into time-reversal symmetry and quantum entanglement in chaotic systems, providing insights into the thermalization processes and information scrambling in black holes, as suggested by the holographic principle.

The intersection of quantum chaos with quantum information theory introduces another layer of complexity and potential. Pioneering studies suggest that the chaotic behavior of quantum circuits could enhance the capabilities of quantum computers, particularly in tasks involving optimization and searching algorithms. The sensitivity of quantum chaotic systems to initial conditions might be harnessed to amplify quantum effects, potentially improving the efficiency of quantum algorithms.

Furthermore, the conceptual framework of random matrix theory continues to be a powerful tool in understanding the spectral properties of quantum chaotic systems. Recent theoretical developments have expanded its application into more complex systems, including those exhibiting non-Hermitian physics, where the interplay between chaos and exceptional points (non-Hermitian degeneracies) has unveiled new chaotic dynamics.

As we delve deeper into the quantum realm, the fusion of chaos theory with quantum mechanics not only challenges our fundamental understanding of nature but also holds promise for innovative technological advancements. By dissecting the chaotic aspects of quantum systems, researchers are not only questioning the very fabric of reality but are also paving the way for potential applications ranging from secure quantum communications to ultra-sensitive measurement devices, exploiting chaos for enhanced performance.

10.7 Challenges and Opportunities for Innovation

The nuanced terrain of chaos theory not only presents various challenges but also opens up a plethora of opportunities for innovation. The delicate balance between the inherently unpredictable nature of chaotic systems and the quest for precise mathematical descriptions provides a fertile ground for substantial breakthroughs in both theoretical and applied sciences.

One prominent challenge in this field is the high sensitivity of chaotic systems to initial conditions, often referred to as the butterfly effect. This sensitivity means that even minuscule differences in starting conditions can lead to vastly different outcomes, making long-term prediction and control of chaotic systems inherently problematic. Addressing this challenge requires innovative approaches to prediction modeling and simulation. Here, machine learning and artificial intelligence have started to play a critical role. Techniques such as deep learning have shown potential in recognizing patterns within chaotic systems that are beyond human discernment.

10.7. CHALLENGES AND OPPORTUNITIES FOR INNOVATION

Moreover, the integration of chaos theory in real-world applications introduces new complexities. While the theoretical aspects are evolving, applying these theories practically — such as in climate modeling, stock market algorithms, and engineering systems — requires innovation in translating complex mathematical concepts into applicable solutions. This translation is not straightforward because practical implementations must consider noise, non-idealities, and computational limitations.

To visualize the challenge, consider the problem of modeling financial markets using chaos theory. Traditional models often fail to predict sudden market crashes or extreme events, which are essentially chaotic in nature. A promising approach employs hybrid models that combine traditional financial theories with chaotic mathematics, potentially offering a more robust framework for understanding market dynamics. Here, a diagram of a hybrid model structure could illustrate how components of standard financial forecasts can be integrated with chaotic predictors for volatility and cyclic behaviors.

Another significant challenge is in education and interdisciplinary collaboration. As chaos theory applications spread across different fields, the need for scientists who can understand and apply these concepts across boundaries grows. Developing curricula that include chaos theory in disciplines like biology, economics, and engineering poses challenges in terms of both content creation and delivery methods.

Conversely, these educational and interdisciplinary challenges also create unique opportunities. For instance, the development of universal chaos theory frameworks that can be easily adapted for multiple disciplines could revolutionize how educational institutions approach complex systems training. Interdisciplinary research teams can foster innovation by combining insights from different fields, leading to novel solutions that a single-discipline perspective might miss.

On the technology front, the exponential growth in computational power and data collection technologies offers unprecedented opportunities for chaos theory research. Improved computational capabilities allow for more detailed simulations and analyses of chaotic systems, which were previously impossible due to hardware constraints. Coupled with big data, researchers can uncover new patterns in chaotic systems by analyzing large datasets that were previously unmanageable, providing deeper insights into the behavior of these complex systems.

Interactive visualizations and simulations are crucial tools for both research and communication in chaos theory. Employing advanced graphics capabilities can make these abstract concepts more tangible and understandable, facilitating better communication between scientists and the public and within the scientific community itself. For example, the use of dynamic plotting tools built with `pgfplots` and `tikz` in LaTeX enables the visual representation of chaotic attractors and their complexities.

In summary, while challenges abound in understanding and managing chaotic systems, each obstacle also presents a unique opportunity for scientific and technological advancement. By embracing these challenges, the scientific community can drive forward the boundaries of what is possible in chaos theory and its applications, ultimately harnessing chaos for practical innovation.

10.8 Ethics and Implications of Advanced Chaotic Systems

The deployment and technological incorporation of advanced chaotic systems invariably raise significant ethical questions and societal implications. As chaos theory further ingrates itself into various sectors including finance, defense, health, and artificial intelligence, we must consider the potential consequences of these applications, not only in terms of technological impact but also in their broader ethical dimensions.

Consider the realm of predictive analytics, a field growing symbiotically with advances in chaos theory. Financial institutions leveraging chaotic models to predict stock market trends could potentially influence market behavior in unpredictable ways, potentially leading to market instabilities. Here, the ethical dimension revolves around the accountability of using such predictive technologies. Who is responsible when a prediction fails spectacularly? How much transparency should be required in these models? The complexity inherent in chaotic systems makes these questions particularly poignant, highlighting the need for rigorous ethical frameworks that consider both the unpredictability and the butterfly-like sensitive dependence inherent in chaotic systems.

In the domain of national defense, chaotic models are integral to developing simulations and strategic assessments that could inform decision-making processes about military operations. The use of chaotic dynamics in drone warfare and surveillance systems also introduces ethical challenges that hinge on the balance between national security and privacy rights. The opacity of highly intricate chaotic algorithms could make it difficult for even

CHAPTER 10. THE FUTURE OF CHAOS: EMERGING TRENDS AND THEORIES

their developers to predict or understand the outcomes of their use, hence complicating the ethical landscape. Ensuring that these technologies are not only effective but also justifiable in their applications requires ongoing ethical scrutiny and adaptive legislative frameworks that can keep pace with technological advancement and the intricacies of chaos theory.

Healthcare is another critical area where chaos theory's application is burgeoning, particularly through its integration in genetic algorithms and personalized medicine. The capability to predict disease outcomes based on chaotic models presents immense implications for patient treatment plans and health outcomes. However, this also raises substantial concerns regarding data privacy, the potential for bias in algorithmic formulations, and the ethical handling of genetic information. It becomes imperative to develop guidelines that safeguard patient data and ensure equitable use of these advanced systems across different populations.

Turning to artificial intelligence, the integration of chaotic dynamics in neural networks and machine learning algorithms can enhance their ability to learn from complex patterns and improve decision-making processes. This presents significant benefits for optimizing logistic operations, managing large-scale network systems, or automating certain types of cognitive or physical labor. Yet, as AI systems increasingly influence every aspect of human life, from job markets to personal relationships, ensuring they operate under ethical principles that prevent misuse and promote transparency and fairness is crucial.

As we continue to explore the vast potential of chaos theory in various applications, creating platforms for ongoing dialogue between technologists, ethicists, and policymakers is vital. This will facilitate not only the development of regulations that are in tune with both technological possibilities and societal norms but also promote an informed public discourse on what it means to live alongside increasingly sophisticated technologies.

Thus, we see that the adoption of advanced chaotic systems must go hand-in-hand with thoughtful ethical consideration and robust policies that comprehend the far-reaching impact these technologies might have on various aspects of human life. By fostering an environment where innovation is paired reliably with responsibility, chaos theory can be implemented in a manner that respects both its power and its complexities.

www.ingramcontent.com/pod-product-compliance
Lightning Source LLC
Chambersburg PA
CBHW052150220526
45471CB00004B/1610